BORN IN AFRICA

Also by Martin Meredith

The Past Is Another Country:
Rhodesia – UDI to Zimbabwe

The First Dance of Freedom:
Black Africa in the Postwar Era

In the Name of Apartheid:
South Africa in the Postwar Era

South Africa's New Era:
The 1994 Election

Coming to Terms:
South Africa's Search for Truth

Africa's Elephant:
A Biography

Mugabe:
Power, Plunder and the Struggle for Zimbabwe

Fischer's Choice:
A Life of Bram Fischer

The State of Africa:
A History of the Continent Since Independence

Diamonds, Gold and War:
The Making of South Africa

BORN IN AFRICA

THE QUEST FOR THE ORIGINS OF HUMAN LIFE

MARTIN MEREDITH

SIMON & SCHUSTER

London · New York · Sydney · Toronto · New Delhi

A CBS COMPANY

First published in Great Britain by Simon & Schuster UK Ltd, 2011
This edition published in Great Britain by Simon & Schuster UK Ltd, 2012
A CBS COMPANY

1 3 5 7 9 10 8 6 4 2

Simon & Schuster UK Ltd
1st Floor
222 Gray's Inn Road
London WC1X 8HB

www.simonandschuster.co.uk

Simon & Schuster Australia, Sydney
Simon & Schuster India, New Delhi

PICTURE CREDITS
2, 8, 13, 21, 22, 27, 34, 37, 38, 41 © Corbis
3, 6, 7, 12, 18, 19, 24, 25, 26, 29, 30, 31, 32, 33 © Science Photo Library
1, 9, 10, 11, 14, 15, 16, 17, 28, 35, 40, 42 © Getty Images
23, 36 © David L. Brill/Brill Atlanta
5, 39 © University of the Witwatersrand Archives Johannesburg
20 © Pennsylvania State University
4, p.d. (E.H.P.A. Haeckel, The Evolution of Life, 1879)

A CIP catalogue record for this book is available from the British Library.

ISBN 978-1-84739-276-3

Designed by Brent Wilcox
Typeset in Adobe Garamond
Printed and bound by CPI Group (UK) Ltd, Croydon, CR0 4YY

We carry within us the wonders
we seek without us: there is all
Africa and her prodigies in us.

SIR THOMAS BROWNE
Religio Medici (1643)

Contents

PART TWO

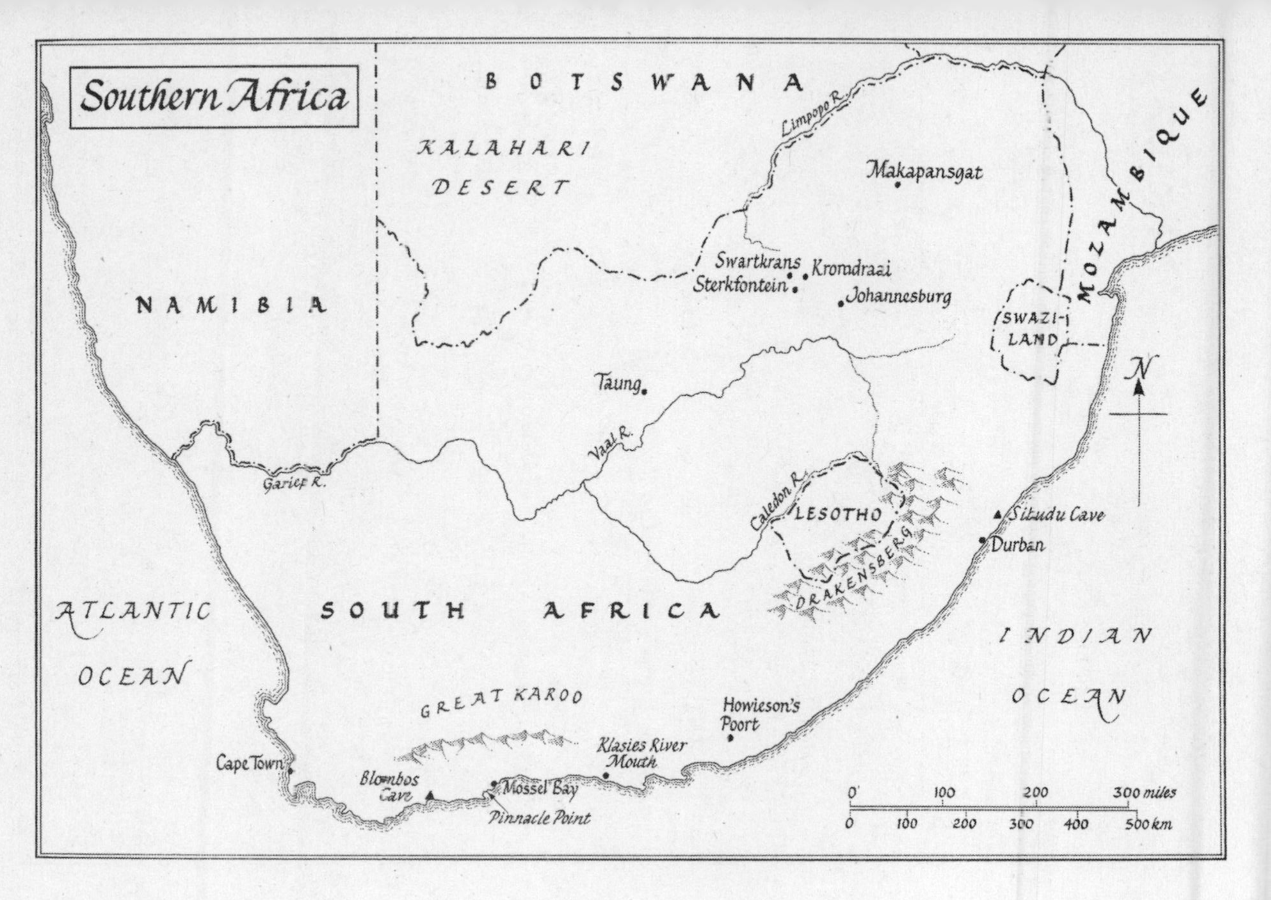
Southern Africa
BOTSWANA
KALAHARI DESERT
NAMIBIA
MOZAMBIQUE
Limpopo R.
Makapansgat
Swartkrans
Kromdraai
Sterkfontein
Johannesburg
SWAZI-LAND
Taung
Vaal R.
Gariep R.
Caledon R.
LESOTHO
DRAKENSBERG
Sibudu Cave
Durban
N
ATLANTIC OCEAN
SOUTH AFRICA
INDIAN OCEAN
GREAT KAROO
Howieson's Poort
Klasies River Mouth
Cape Town
Blombos Cave
Mossel Bay
Pinnacle Point
0 100 200 300 miles
0 100 200 300 400 500 km

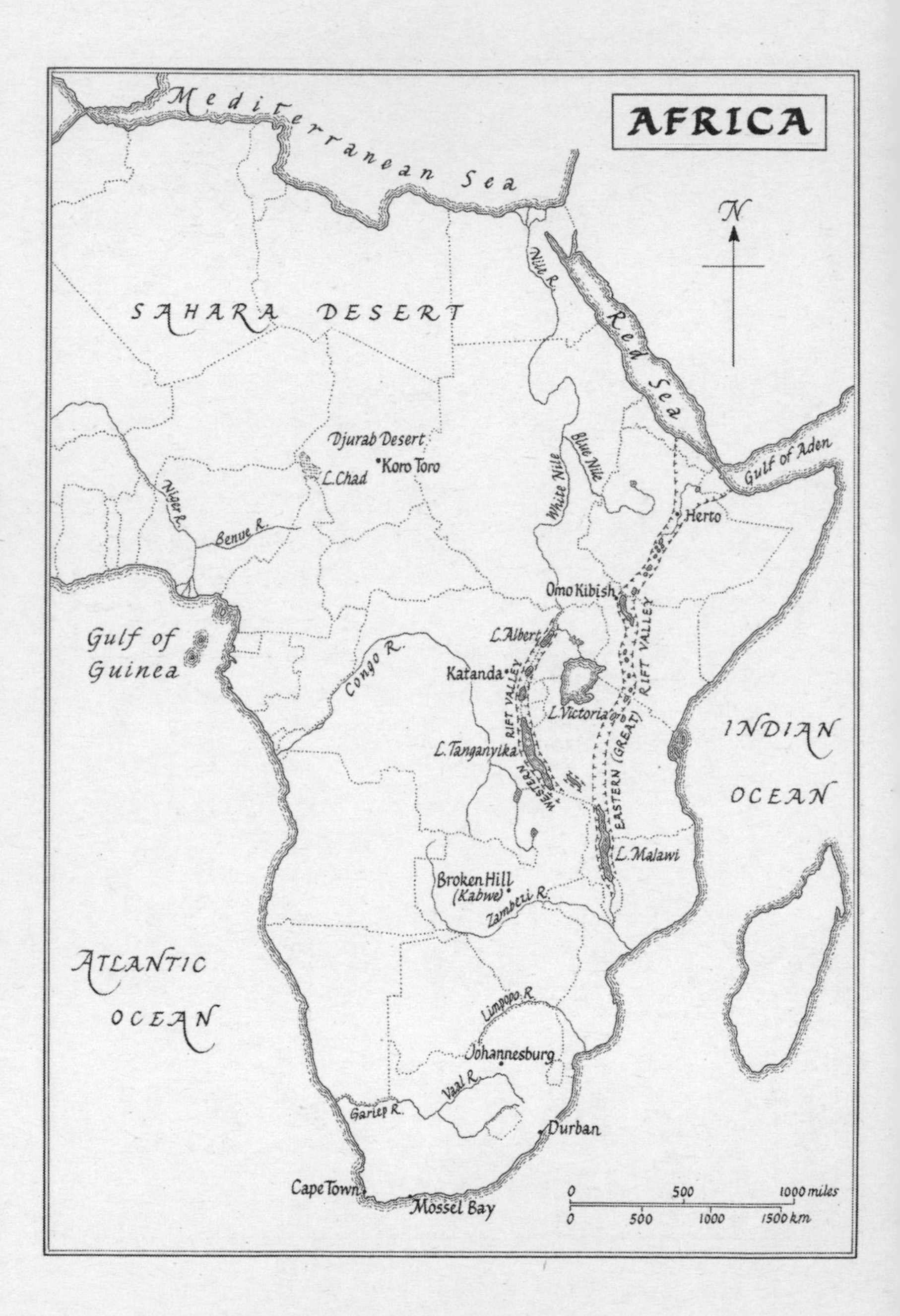
AFRICA
Mediterranean Sea
N
SAHARA DESERT
Nile R.
Red Sea
Djurab Desert
Koro Toro
L. Chad
Gulf of Aden
Blue Nile
White Nile
Niger R.
Benue R.
Herto
Omo Kibish
L. Albert
Gulf of Guinea
Congo R.
Katanda
RIFT VALLEY
L. Victoria
INDIAN OCEAN
L. Tanganyika
WESTERN RIFT VALLEY
EASTERN (GREAT) RIFT VALLEY
L. Malawi
Broken Hill (Kabwe)
Zambezi R.
ATLANTIC OCEAN
Limpopo R.
Johannesburg
Vaal R.
Gariep R.
Durban
Cape Town
Mossel Bay
0 500 1000 miles
0 500 1000 1500 km

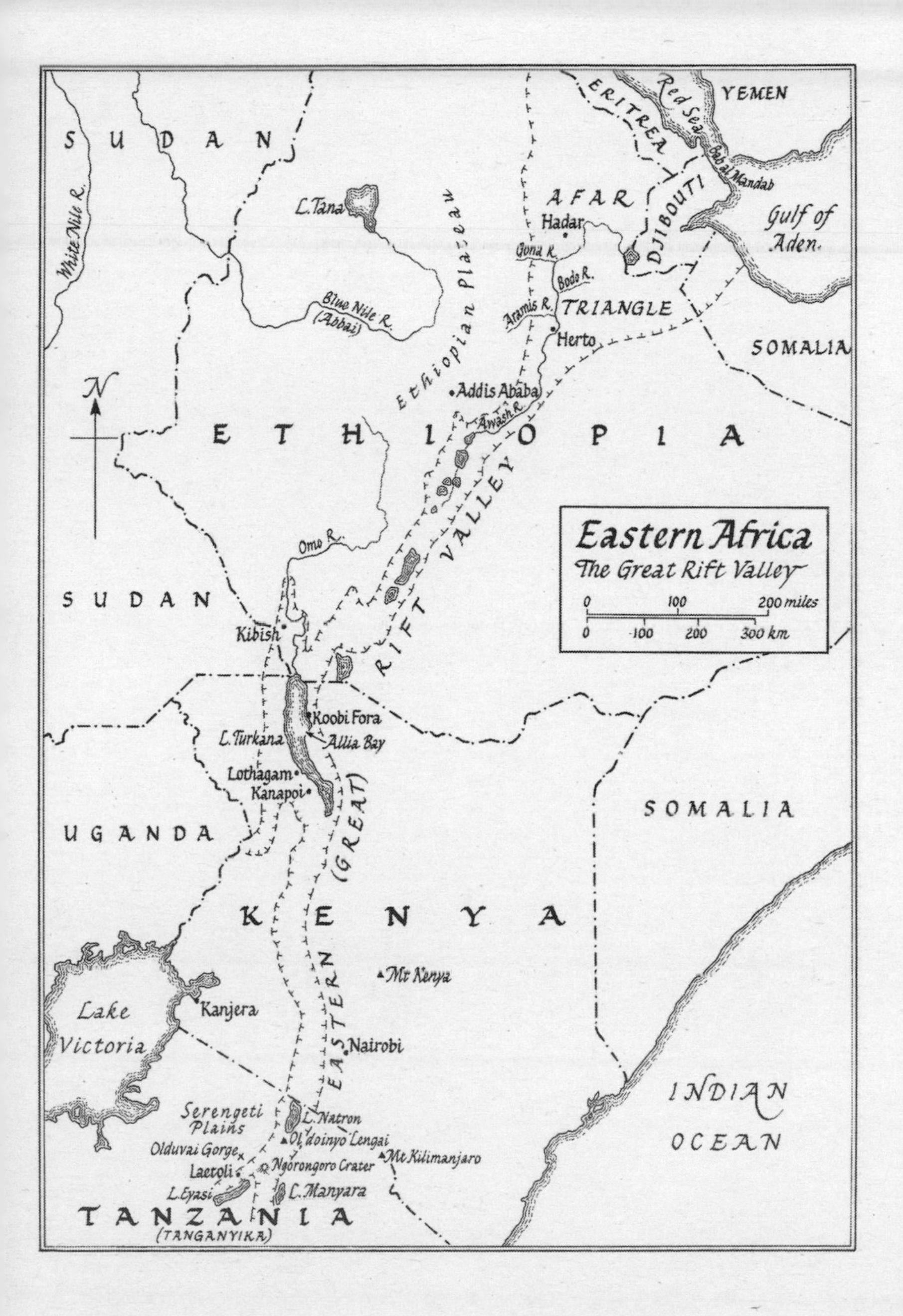

Eastern Africa
The Great Rift Valley
0
100
200 miles
0
100
200
300 km
SUDAN
White Nile R.
L. Tana
Blue Nile R. (Abbai)
Ethiopian Plateau
ERITREA
Red Sea
YEMEN
Bab al Mandab
AFAR
Hadar
Gona R.
Bodo R.
Aramis R.
TRIANGLE
Herto
DJIBOUTI
Gulf of Aden.
SOMALIA
Addis Ababa
Awash R.
N
ETHIOPIA
RIFT VALLEY
Omo R.
SUDAN
Kibish
Koobi Fora
L. Turkana
Allia Bay
Lothagam
Kanapoi
(GREAT)
SOMALIA
UGANDA
KENYA
EASTERN
Mt Kenya
Lake Victoria
Kanjera
Nairobi
INDIAN OCEAN
Serengeti Plains
L. Natron
Ol doinyo Lengai
Olduvai Gorge
Mt Kilimanjaro
Laetoli
Ngorongoro Crater
L. Eyasi
L. Manyara
TANZANIA
(TANGANYIKA)

Preface

Africa does not give up its secrets easily. Buried there lie answers to our questions about the origins of humankind. After a century of investigation, scientists have transformed our understanding of the beginnings of human life. Many remarkable discoveries have been made. Yet even as the evidence about human evolution has continued to grow, so the riddle has become ever more complex. And ultimate clues still remain hidden.

This book follows the endeavours of scientists striving to uncover the mysteries of human origins over the past 100 years. The obstacles they faced have been formidable. Some 7 million years have passed since the precursors of humankind began to evolve in Africa. The only signs of their existence are fossil remains concealed in a landscape that has changed dramatically in that time. Whole parts of the continent have been raised by tectonic upheavals; mountains have been thrust upwards; lakes and rivers have come and gone; erupting volcanoes have covered great swathes of land under layers of lava and ash.

The route back to this ancient world has been marked by misfortune, false hopes, fraud and extraordinary feats of skill and endurance. The early stages of the quest were dominated by a handful of ambitious individuals, obsessed by their work and driven by hopes of fame and glory. Their goal was to find the oldest human ancestor. Each discovery they made was acclaimed as having iconic significance. From the outset, however, the science of palaeoanthropology has been renowned not just for the exploits of researchers in the field but for their intense rivalry, personal feuds and fierce controversies. One field scientist observed ruefully in his memoirs how the profession was plagued by 'treachery, cutthroat competition and backstabbing'.

In more recent times, a host of other scientists—molecular biologists, biochemists, geneticists, palaeoclimatologists, geochronologists—have played an increasingly influential role in this giant detective saga. The focus of attention has broadened to include the search for the origins of modern humans as well as human ancestors. New controversies have erupted. Rival schools of thought have fought each other as tenaciously as in the past.

The results of the quest have been momentous. Scientists have identified more than twenty species of extinct humans. They have firmly established Africa as the birthplace not only of humankind but also of modern humans. They have revealed how early technology, language ability and artistic endeavour all originated in Africa; and they have shown how small groups of Africans, possessing new skills, spread out from Africa in an exodus 60,000 years ago to populate the rest of the world.

We have all inherited an African past.

The first part of this book focuses upon the exploits of key field scientists, starting with the pioneer researchers of the early twentieth century. Their task was not only to find significant fossils—the principal evidence of human evolution—but to convince a sceptical scientific establishment of the importance of their discoveries. Some fossil finds remained in dispute for years. Modern researchers pushing back the frontier of human origins to 7 million years ago have encountered similar hurdles.

The second part of the book opens at that primordial frontier and moves forward along the trail of discoveries leading to the emergence of our own species, *Homo sapiens*, and its gradual migration around the world. What stands out is not only the remarkable range of scientific discoveries that have been made but the extent of the vast hinterland that remains to be discovered.

Introduction

While working on his revolutionary theories about evolution, the naturalist Charles Darwin concluded that the most likely birthplace of humankind was Africa, since it was the homeland of gorillas and chimpanzees, apes which he deemed to be our closest living relatives. Humans and apes, said Darwin, had probably shared a common ancestor in Africa.

'In each great region of the world', he wrote in *The Descent of Man*, published in 1871, 'the living mammals are closely related to the extinct species of the same region. It is therefore probable that Africa was formerly inhabited by extinct apes closely allied to the gorilla and chimpanzee; and as these two species are now man's nearest allies, it is somewhat more probable that our early progenitors lived on the African continent than elsewhere'.

The idea that humans were related to an African ape caused uproar in Victorian England. 'Descended from the apes!' exclaimed a bishop's wife to her husband. 'My dear, let us hope that it is not true, but if it is let us pray that it will not become generally known'.

The Victorian era was accustomed to Christian doctrines about life on earth that regarded humans as unique, a special creation separate from the rest of the animal world, made in the image of God and given dominion over nature. What the public found so offensive was not the general theory of evolution that Darwin propounded. Geologists had already shown that the earth was far older than allowed for in the Book of Genesis and that it had changed significantly over a vast period of time. Archaeologists had found stone tools alongside extinct animals from the Ice Ages indicating that humans, too, had been on earth for far longer than the 6,000 years laid down by biblical chronology.

Victorian society was ready to accept the idea of a changing world. Evolution could be seen as the gradual unfolding of a divine plan. It represented progress—the constant improvement of form and function—a subject of immense appeal to Victorian audiences. Humans, standing atop the ladder of evolution, were clearly life's supreme refinement. Indeed, evolution, it was said, had been planned by a wise and benevolent God to result in human life. Darwin's theory of common descent—the proposition that all living things were descended from a common ancestry—was swiftly accepted.

Other aspects of Darwin's explanations about life on earth, however, caused endless trouble, not only with the public but among the scientific community. Evolution, Darwin maintained, did not rely on any supernatural power. It was governed solely by the response of a species to its physical and biological environment. Every species produced more offspring than could survive from generation to generation. In 'the struggle for existence', said Darwin, it was those individuals best able to adapt to the demands of the prevailing environment—'the fittest'—that would survive. The traits or variations that enabled them to adapt would be more prevalent in the next generation. Adaptation was thus the driving force behind evolution. By a process of natural selection, the less fit were eliminated. 'Common descent with modification' was the framework for understanding the history of life. Over an immense period of time, infinitesimal changes wrought by the struggle for survival had led to the evolution of species. This process of natural selection applied to all life on earth—including humans. Darwin treated humankind as just one species among all others, moulded by the same evolutionary forces.

The implications of Darwin's theory were profound. It opened up the possibility of a world without purpose, or direction, or long-term goal, a world that seemed to be no more than a product of chance. It stripped humankind of its unique status and was seen to undermine Victorian respect for hierarchy and social order. Above all, it threatened the very foundations of Christian belief and morality. On one of his visits to the British Museum, Darwin was pointed out by a clergyman as 'the most dangerous man in England'.

Even Darwin's scientific colleagues found difficulty in accepting some of his ideas, especially his emphasis on natural selection as being the mechanism of change. One eminent scientist dismissed natural selection as the 'law of higgledy-piggledy'. Most scientists disliked the idea that evolution could be an open-ended process of adaptation and divergence. They preferred to believe that evolution was guided inexorably in the direction of progress towards humankind.

Nor did they concur with Darwin's model of an evolutionary tree with numerous branches to explain the extent of biological diversity. The model they favoured was based on linear development—a tree of life with a main trunk leading upwards from 'lower' organisms at the bottom to humans at the top—a modernised version of the ancient notion of a Chain of Being that had previously been used to explain life on earth. They remained convinced that evolution was all part of a purposeful process, directed towards a predetermined goal.

There was also disagreement about the way that human faculties were said to have developed. Darwin speculated that human ancestors had moved from a forest environment onto the open plains of Africa, acquiring the ability to walk upright on two legs as a better means of locomotion. Bipedal locomotion had thus been the key breakthrough—the first attribute separating human ancestors from the ape masses. It had freed their hands for primitive toolmaking, which in turn had stimulated the growth of their intelligence. Other apes meanwhile had stayed in the trees, continuing to use their hands as a means of locomotion; they had consequently never acquired the need for additional intelligence. In other words, Darwin regarded the development of higher human faculties as no more than a by-product of a change in the mode of locomotion by one particular group of African apes.

The theory supported by most other scientists was that the brain had been the original driving force behind human evolution. Impressed by the large size of the modern human brain, they believed that it must have been sheer brainpower that had propelled humans along the road to preeminent status.

Darwin's suggestion that Africa was the cradle of humankind was also challenged. An influential German biologist, Ernst Haeckel, argued that Asian apes—orang-utans and gibbons—were more closely related to humans than African apes were, making Asia a more likely birthplace. Haeckel's tree of life also differed from Darwin's. He proposed the existence of an intermediate link between humans and apes that he called 'Ape-like Man', or *Pithecanthropus.* His reasoning was that the human capacity for speech must have required more than a single evolutionary step in which to develop. Haeckel described this hypothetical link as a hairy, primitive creature with a long skull and protruding teeth that walked semi-erect. This idea of an intermediate figure from the past became popularly known as 'the missing link'.

Whatever theories scientists chose to air, there was scant evidence on the ground to support any of them. Only one possible candidate for the missing link had come to light: parts of a skeleton unearthed in 1856 by quarry workers clearing out a limestone cave in the Neander Valley near Düsseldorf in Germany. The remains—the top of a cranium, some leg and arm bones—belonged to an individual who was evidently human but unlike any other human known. The individual was heavily built, short in stature, with prominent ridges above the eyes and a low, receding forehead—similar to an ape but with a modern-sized brain.

The reaction of scientists to this discovery was mixed. The biologist Thomas Huxley emphasised the importance of brain size as a defining characteristic of humans, setting a standard that was to be followed by subsequent generations of researchers. He concluded therefore that although the specimen had some apelike features, it was nevertheless fully human. Others argued that it was simply a deformed or diseased human, perhaps an idiot or a wild man.

But the Irish anatomist William King considered it to be distinctly different from modern humans, and in 1864 he accorded it the species name *Homo neanderthalensis*—Neanderthal Man—making the first formal recognition that another human species other than *Homo sapiens* had existed on earth. As the new science of palaeoanthropology developed over the years, it was to become a common feature that scientists examining the same evidence reached diametrically opposed conclusions.

The next significant discovery was made in southeast Asia. Inspired by Ernst Haeckel's suggestion that the 'missing link' would be found in Asia, a young Dutch anatomist, Eugène Dubois, travelled there in 1887 on assignment as an army doctor hoping to find *Pithecanthropus*. After two unsuccessful years on Sumatra, he moved to the island of Java. In 1891, his team of labourers found a hominid molar, then a skullcap, at a site near the village of Trinil. The skullcap had strong brow ridges and no forehead, similar to a male ape, but a brain size that was large for an ape though small for a human. The following year Dubois's team found a left femur, or thighbone, that was humanlike in both size and shape, indicating an upright posture. Dubois claimed that all three remains belonged to the same individual. He was convinced he had found the missing link and

See also p. 114

named it *Pithecanthropus erectus*—'upright ape-man'. To the world at large it became known as Java Man.

Returning to Europe in 1895, Dubois was acclaimed for his exploits in Java but, to his dismay, he found most of the scientific community sceptical about his conclusions. His monograph on the subject was openly mocked. One prominent German anatomist, Rudolph Virchow, declared that the Java bones belonged to a giant gibbon. A rising young Scottish anatomist, Arthur Keith, took a different approach. Following Huxley's lead, he argued that brain size was the determining factor. Java Man, therefore, was neither an ape nor an intermediate link between apes and humans, as Dubois had claimed, but a primitive human; its brain size was estimated to be 900 cubic centimetres, about two-thirds the size of a modern human brain; it therefore crossed the threshold—'the cerebral Rubicon'—that qualified it to be classified as a human, albeit one with low intelligence, lower than Neanderthal Man. Keith concluded that the Java bones fitted neatly into a ladderlike progression in the human line: first came *Pithecanthropus* with its comparatively small brain; then Neanderthal Man with its bigger brain; and finally, true man.

By the end of the nineteenth century, nearly eighty scientific books and articles had appeared discussing Dubois's *Pithecanthropus*, almost all of them disagreeing with his claims. Embittered by the reaction of his colleagues, Dubois withdrew from scientific debate and hid his Java bones inside cabinets in his dining room, refusing to let any researchers see them.

Meanwhile, the fortunes of Neanderthal Man as a contender for the missing link had improved. Further discoveries of Neanderthal fossils were made in Belgium, Croatia, Germany and France, demonstrating conclusively that the Neander Valley specimen was not an aberration. But the reputation of Neanderthals was soon ruined. In 1908, two young priests excavating a small cave near the village of La

Chapelle-aux-Saints in central France, unearthed the most complete Neanderthal skeleton yet discovered. The skeleton was sent for examination to Marcellin Boule, a renowned palaeontologist at the Muséum National d'Histoire Naturelle in Paris. Boule's verdict was decisive. He described the Chapelle-aux-Saints individual as a coarse brute with a short, thick-set body, heavy overhanging eyebrow ridges, retreating forehead, stooped posture, bent knees and low intelligence. An illustration authorised by Boule to depict his findings was published in 1909 by the French magazine *L'Illustration* and by the *Illustrated London News*. It portrayed an excessively hairy, apelike thug wielding a club, teeth bared, eyes glaring—an image that became embedded in popular culture for more than fifty years. Boule excluded all possibility that Neanderthals could have stood in the direct line of human descent; they were an unfortunate offshoot. Other leading scientists—including Arthur Keith—concurred. Once again the missing link was missing.

The gap was soon filled by one of the most audacious hoaxes in history—a fossil find that fooled the British scientific establishment for more than forty years. In 1908, it was said, a labourer digging in a gravel pit at Piltdown in southern England found fragments of thick human skull which he passed to Charles Dawson, a local lawyer and amateur fossil hunter. Over the next few years Dawson visited the site frequently, and in 1911 he found another fragment from the same skull. He took his finds to Arthur Smith Woodward, the Keeper of Geology at the British Museum and an eminent palaeontologist, who expressed keen interest in them. In 1912, Woodward set off for a summer of digging at Piltdown, joining Dawson and a French palaeontologist, Father Pierre Teilhard de Chardin. By the end of the season they had found three more pieces of skull bone along with an apelike jaw, assorted mammal fossils and a few crude stone tools—altogether a remarkable haul.

Back in London, Woodward pieced together the bits and pieces to produce a reconstruction of the skull; and in December 1912, he unveiled it at a crucial meeting of the Geological Society, naming it *Eoanthropus dawsoni*—'Dawson's dawn man'.

The skull accorded neatly with the prevailing view among scientists about what a proto-human should look like: it possessed a relatively large brain while retaining certain apelike features such as the jaw. A leading neuroanatomist, Grafton Elliot Smith, who specialised in brain studies and supported the 'brain-led-the-way' school, concurred with Woodward that *Eoanthropus dawsoni*—or Piltdown Man, as it was popularly known—represented the ancestor of modern humans. Arthur Keith initially had some doubts, but after further discoveries of a canine tooth in the gravel pit at Piltdown and fragments of a second individual at another site two miles away, he too fell for the hoax.

Much to the satisfaction of British scientists, Piltdown Man put England firmly on the anthropological map, trumping French and German claims. Indeed, it became a matter of national pride that the earliest human ancestor had been found on home soil. News of the discovery swept around the world.

No one, meanwhile, gave much thought to Darwin's suggestion forty years before that Africa was the most likely birthplace of humankind.

PART ONE

CHAPTER I

The Valley of Wild Sisal

Setting out on foot across the Maasai Steppe in 1913 at the head of a column of fifty porters, Hans Reck, a twenty-seven-year-old German geologist, had no clear idea how to find his destination. Behind him rose the snow-capped peak of Kilimanjaro, the highest mountain in Africa. Ahead lay volcanic highlands lining the Great Rift Valley. Reck's mission was to investigate a ravine to the west of an extinct volcano named Ngorongoro that had aroused keen interest in Berlin. But he had been given only vague instructions about its location.

Two years before, a German entomologist, Professor Wilhelm Kattwinkel, had stumbled across the ravine by chance while leading a medical expedition to this remote part of what was then German East Africa (now Tanzania). When Kattwinkel had asked local Maasai tribesmen the name of the ravine, they had thought he was referring to the wild sisal growing there—*Sansevieria ehrenbergii*—and had told him they called it 'oldupai'. Kattwinkel had duly recorded the name of the ravine in German as 'Oldoway'.

Exploring the eroded slopes of Oldoway, he had made a small collection of ancient fossils, taking them back with him to Berlin. When it was realised that some of the bones belonged to an unknown species of three-toed horse, there was considerable excitement. With the personal support of the kaiser, a new expedition had been authorised under the auspices of the Universities of Berlin and Munich, and Hans Reck had been chosen to lead it.

A specialist in volcanology, Reck had already proved his ability to handle African expeditions. In 1912, he had been appointed leader of a university expedition to Tendaguru in the southern region of German East Africa which had uncovered a complete skeleton of a *Braciosaurus*, one of the largest land animals ever to have lived. His team of porters from Tendaguru had joined him for this new expedition. But the journey across the Maasai Steppe proved arduous: In the searing heat Reck's column struggled to keep pace and was strung out for miles. Water was scarce.

Climbing up the western escarpment of the Great Rift Valley onto the slopes leading to Ngorongoro, Reck caught sight in the distance of Ol doinyo Lengai, the Maasai's 'Mountain of God', so named because it was still an active volcano. Reaching the rim of Ngorongoro, he marvelled at the spectacle around him: Below lay a collapsed crater, twelve miles across and 2,000 feet deep, forming a natural amphitheatre of 100 square miles that teemed with wildlife. Once rising to a height of 15,000 feet, the Ngorongoro volcano had exploded 2.5 million years ago, causing its dome to crumple inwards and leaving walls that were only half their original size.

But Reck was still unable to find out anything about Oldoway. A German official he met at Ngorongoro had never heard of it. Local Maasai also had no information, but they agreed to show him the way to a spring called Langavata on the western side of Ngorongoro which overlooked the great Serengeti Plains stretching away to the horizon.

For three days, Reck wandered along the fringes of the Serengeti. Then on 7 October 1913, he set up camp on the rim of a steep gorge. The surrounding terrain seemed familiar. Looking at photographs that Professor Kattwinkel had taken two years previously, Reck realised that he had found Oldoway.

In the months that followed, Reck collected more than 1,700 fossils and completed a geological survey of the area. He discovered that the gorge offered a remarkable geological record of past millennia. Its walls consisted of five distinct layers, or 'beds', of lava and ash, providing a sequence of time dating far back into antiquity. At the base was a layer of black lava; above stood a layer cake of colours—rich copper sandwiched between duller buffs and greys. No means of accurate dating were then available. But in time Oldoway was to yield crucial clues about the importance of the volcanic regions of the Great Rift Valley in revealing the origins of humankind.

The Great Rift Valley acts as a history book of the deep past. Over the last 10 million years, as two of Africa's tectonic plates have slowly pulled apart, a giant fissure has developed in the earth's crust, running for more than 3,000 miles from the lower reaches of the Zambezi Valley in Mozambique, through eastern Africa to the Red Sea where it divides Africa from Arabia. Tectonic upheavals and volcanic eruptions have transformed a relatively flat rain-forest region into a dramatic landscape of mountains, lakes and a complex array of fractures, faults and scarps. Ethiopia's landmass rose 8,000 feet above the surrounding plains, like a huge blister on the continent's skin, to form the largest volcanic massif in Africa. Eruptions in Kenya built a similar dome. Most of the uplift occurred after 7 million years ago. To the west, running parallel to the main Rift Valley, a new fissure began to develop, creating another chain of mountains and lakes, including Lake Tanganyika, the deepest lake in Africa, where the lake bottom lies 2,200 feet below sea level. Shoulders of land along the western rift were pushed up to form new mountain ranges such as the Rwenzori massif, the fabled Mountains of the Moon once thought by ancient Greek geographers to be the source of the Nile. Many of the ancient lake basins along the Rift have since disappeared, buried under layers of lava, ash, sand and mud, sediments which have subsequently been thrust upwards by tectonic movements and then exposed by erosion from wind and rain. Among the sediments lie millions of fossils, giving glimpses of life long past.

Oldoway was once part of the shoreline of a shallow alkaline lake formed about 2 million years ago and fed by streams and rivers spilling down from the volcanic highlands to the east and south. Volcanic ash from two active volcanoes—Olmoti and Kerimasi—was periodically deposited on the lake, blown there by the prevailing wind. Over a period of about 400,000 years, the lake gradually shrank and disappeared. In more recent times—about half a million years ago—a seasonal river began to cut its way through the accumulated layers of lake sediments and ash deposits, eventually carving out a steep-sided gorge, with cliffs that in places fell 300 feet. By chance, part of the gorge followed the shoreline of the prehistoric lake, an area rich in ancient fossils, as Hans Reck discovered.

In December 1913, after nearly three months exploring Oldoway, Reck was almost ready to leave when one of his workmen reported finding a human skeleton buried in a crouched position in what was called Bed II, one of the oldest layers in the gorge wall. When he inspected the site, Reck immediately understood its significance. The skeleton clearly belonged to a modern human—*Homo sapiens*—but it lay at a level where extinct Pleistocene animals had been found. If the skeleton was as old as its surroundings, then it meant that it was one of the oldest human finds ever made.

'It is impossible to describe the feelings by this sight', wrote Reck in his account of the expedition. 'Joy, hope, doubt, caution, enthusiasm—all this surged wildly to and fro'. In further excavations of the site, Reck could find no evidence that the body had been buried in a grave dug in more recent times. 'The encasing soil-mass and that of the surrounding area were one and the same material. The geological conclusion became ever clearer that the man, just like the animals, was a contemporary fossil of its stratum and had not been introduced into it only later as a more recent grave'.

Reck returned to Berlin in March 1914 with the skull wrapped up in his personal linen while the skeleton followed with other fossils. An article he wrote in the *Lokal Anzeiger* aroused huge public interest. News of his discovery was also published in London in April by the *Illustrated London News*. In public meetings, Reck argued that 'Oldoway Man' was proof that the human race was 'of considerably greater antiquity than has been imagined'. He speculated that the skeleton was the remains of a man who had drowned 150,000 years ago, challenging the conventional view that modern humans were no more than 100,000 years old at most. But in the controversy that followed, Reck's claims were widely dismissed. Many of his scientific colleagues argued that the body belonged to a Maasai tribesman, buried recently in a much older deposit.

To try to settle the matter, three more German expeditions to Oldoway were launched. Two of them failed to find the gorge. The third managed to reach Oldoway, but no sooner had it arrived than war was declared in August 1914 and its members were ordered to return immediately.

Reck himself stayed in German East Africa after the outbreak of war, working as a government geologist. But in 1916, he was taken prisoner and spent the rest of the war interned in Egypt. Convinced that he had found one of the earliest examples of modern humans, he yearned to return there. However, after the war German East Africa became British territory. Under British rule, it was renamed Tanganyika, and the spelling of Oldoway was changed to Olduvai. The mystery of Oldoway Man stood unresolved. Although Reck managed to return to Olduvai in 1931, its place as a vital link to the human past was to remain hidden for decades longer.

Meanwhile, other discoveries in Africa had taken the limelight.

CHAPTER 2

Dart's Child

ARRIVING IN JOHANNESBURG in January 1923 to take up a post at the new University of the Witwatersrand, Raymond Dart, a twenty-nine-year-old Australian, felt a profound sense of foreboding. Only with great reluctance had he been persuaded to forsake his London-based career for the backwaters of South Africa. In London, as a senior demonstrator in the anatomy department at University College, he had been able to work alongside some of the giants of British medicine; but Johannesburg was little more than an overgrown mining camp, remote from the forefront of medical research. Although given the rank of professor, Dart feared he had taken a wrong turn. Johannesburg, he later recalled, seemed more like a place of exile than one of opportunity.

It was worse than he had expected. He took an instant dislike to Johannesburg, with its endless rows of red-painted, corrugated-iron-roofed buildings. 'It seemed to have progressed little since the days of the gold rush towards the end of the [nineteenth] century and one felt that if a financial slump hit the place, it would become a deserted ghost-town in a matter of days'.

Moreover, he found the facilities offered by the new university to be entirely inadequate. The medical school—a double-storeyed building hidden behind ten-foot-high garrison walls—stood amidst high grass and weeds and exuded 'a general air of dereliction'. The anatomy department consisted of a dissecting hall with three side-rooms, a lecture theatre and an underground basement mortuary, bereft of almost all equipment. He recalled: 'The architect had overlooked the necessity for planning water taps, electric plugs, gas or compressed air for student laboratories'. The walls of the dissecting hall were spattered with dirt and other marks indicating their use for football and tennis practice. On trestle-type dissecting tables lay dried-up portions of corpses covered only by scant hessian sheets. During a preliminary tour of inspection, Dart's American wife, Dora, a former medical student from Cincinnati, was so distressed by the conditions that she burst into tears. To add to his consternation, Dart next discovered that the medical school did not even possess a library.

Nor did he receive much of a welcome from either university colleagues or students. His predecessor had been a popular figure who had been forced to resign, amidst a storm of protest and controversy, as a result of his affair with the chief college typist and his subsequent divorce.

Dart also encountered lingering resentment over his Australian nationality. Australians were disliked by many Afrikaners because of their involvement in the Anglo-Boer war on the side of the British. Shortly before he left London, Dart was shown a letter from Professor Jan Hofmeyr, the university's principal, expressing 'regret that the appointee was Australian'.

Close to despair, Dart resolved to press for improvements to the anatomy department. He began to establish a medical library and a specimen collection. He also tried to keep up research he had started in London on the nervous system and the evolution of the brain. But, frustrated by the lack of equipment and scientific literature, he soon found it necessary to divert his attention to other areas, notably anthropology, in which he had previously taken only a passing interest.

As a medical student in Sydney, he had striven in particular to avoid the subject of bones. Now he was obliged to study bones instead of brains. Recalling his early experience at the University of the Witwatersrand, Dart remarked: 'It would be useless to deny that I was unhappy in the first eighteen months'.

The sequence of events that propelled Dart to international fame began in May 1924. On a visit to a limestone quarry at Buxton near the African village of Taung in the northern Cape, a mining company official, E. G. Izod, was shown what looked like a fossilised monkey skull embedded in limestone rock. It had been thrown up with broken limestone during blasting operations and kept in the manager's office at Buxton as a souvenir. Izod decided it would make an excellent paperweight and took it back to his home in Johannesburg. His son thought it might interest some university friends and showed it to one of Dart's students.

When Dart first saw the monkey skull, he realised it was a significant find. Within minutes he sped off with the skull in his Model T Ford to consult a colleague, Professor R. B. Young, a veteran Scottish geologist. Young was familiar with the geology of the Taung area. By coincidence, he had been commissioned to investigate lime deposits a few miles south of Buxton, and he promised Dart that on a visit he was due to make to the area in November 1924, he would call at Buxton and look out for further likely specimens.

Shortly before Young arrived at Buxton, an alert quarryman, M. de Bruyn, blasting out a section of rock-face, noticed an unusual shape among the breccia blocks. De Bruyn had previously collected a number of fossilised baboon skulls from the site, but this latest object appeared to be different. He was sufficiently intrigued to take two bone-bearing blocks to the manager's office. They were still there when Young called. He immediately recognised their importance, carried them back to Johannesburg and, on 28 November, drove over to Dart's house.

It was an inopportune moment. The Dart household was in the throes of preparing for a marriage ceremony at the house for two friends, at which Dart was to be best man. His wife, Dora, had made elaborate arrangements. But Dart was transfixed by the fossil blocks that Young had brought him. From his knowledge of brain formation, he instantly discerned part of an ape's skull with distinct hominid features.

> I knew at a glance that what lay in my hands was no ordinary anthropoid brain. Here in lime-consolidated sand was the replica of a brain three times as large as that of a baboon and considerably bigger than that of any adult chimpanzee.

The face remained hidden in the rock. But even without it, Dart knew that aspects of the brain-cast meant that he was on the verge of a remarkable discovery.

> I stood in the shade holding the brain as greedily as any miser hugs his gold, my mind racing ahead. Here, I was certain, was one of the most significant finds ever made in the history of anthropology.
>
> Darwin's largely discredited theory that man's early progenitors probably lived in Africa came back to me. Was I to be the instrument by which his 'missing link' was found?

Engrossed by the rock, Dart ignored his wife's remonstrations to get ready for the marriage ceremony. Only when the bridegroom began tugging on his sleeve did he take notice. 'My God, Ray,' said the bridegroom in an agitated tone, 'You've got to finish dressing immediately—or I'll have to find another best man. The bridal car should be here at any moment'.

For the next three weeks, Dart used every spare moment to patiently chip away the matrix from the skull. He had no previous experience of such a task, nor any colleagues to whom he could turn for advice. Nor could he find any relevant textbooks other than what he had brought from London. Nor did he have any suitable tools. Apart from a hammer and some chisels he purchased from a local hardware store, his most useful implement turned out to be his wife's knitting needles that he kept sharpened to a fine point. Day after day he worked in constant fear that the slightest slip of a chisel might shatter the fossil within.

Two days before Christmas, the rock parted and the face of a child emerged. The large brain that Dart had detected belonged not to an adult hominid but to an infant.

Nothing like it had been discovered before. Dart's fossil consisted of an endocranial cast—a natural mould of the inside of the skull—and a well preserved facial structure including both jaws, all twenty of the milk teeth, and the first of the permanent teeth to erupt, the upper and lower first molars.

Losing no time, he began to prepare a report for publication in the prestigious London science journal *Nature*.

> I was aware of a sense of history for, by the sheerest good luck, I had been given the opportunity to provide what would probably be the ultimate answer in the comparatively modern study of the evolution of man.

What particularly impressed Dart were the humanlike features of the Taung specimen: its high, domed forehead; its lack of eyebrow ridges; its large and rounded eye sockets; its lightly built lower jaw; the small profile of its teeth. Instead of protruding like that of an ape, the face had a flatter appearance. Its age, he calculated, on the basis of its teeth structure, was about the same as a human at a similar stage of development—some six years. Its brain, however, appeared to be relatively small. Dart estimated the skull capacity to be 520 cubic centimetres, bigger than a chimpanzee's but smaller than a gorilla's. He was struck in particular by the position of the foramen magnum, the aperture through which the spinal cord leaves the cranium and enters the spinal column: It was situated at the base of the skull rather than towards the rear, as in the case of quadrupedal apes. This could only mean one thing, Dart surmised: The Taung child must have walked upright, like humans.

Excited by these findings, Dart opened his article for *Nature* with the bold claim that the Taung specimen represented 'an extinct race of apes intermediate between living anthropoids and man'—'a man-like ape', possessing 'humanoid' characteristics. To mark its status, he proposed a new genus and species for it: *Australopithecus africanus*: *austral* from the Latin, meaning 'southern'; *pithecus*, of Greek origin, meaning 'monkey' or 'ape'.

Dart speculated freely about its place in the history of human evolution. Because the Taung child had walked upright on two feet, he said, its hands had been freed to assume 'a higher evolutionary role'. It was able to carry out 'more elaborate, purposeful, and skilled movements' than apes, using its hands as 'organs of offence and defence' and for making tools. Its brain structure not only enabled it to process sight, sound and touch more thoroughly than any ape but indicated that it was within reach of the ability to acquire language.

Dart wrote about the attributes of the new species not in sober scientific prose but with breathless enthusiasm:

> They possessed to a degree unappreciated by the living anthropoids the use of their hands and ears and the consequent faculty of associating with the colour, form, and general appearance of objects,

their weight, texture, resilience and flexibility, as well as the significance of sounds emitted by them. In other words, their eyes saw, their ears heard, and their hands handled objects with great meaning and to fuller purpose than the corresponding organs in recent apes. They had laid down the foundations of that discriminative knowledge of the appearance, feeling, and sound of things that was a necessary milestone in the acquisition of articulate speech.

He speculated, too, about the location where it had been found. Taung, he noted, was on the fringe of the Kalahari Desert. It was some 2,000 miles distant from the luxuriant tropical forests of central Africa—the natural habitat of ape populations. In central Africa, he wrote, 'Nature was supplying with profligate and lavish hand an easy and sluggish solution, by adaptive specialization, of the problem of existence'. But anthropoid groups venturing into southern Africa, where conditions were harsher, had been obliged to develop new techniques. 'For the production of man a different apprenticeship was needed to sharpen the wits and quicken the higher manifestations of intellect—a more open veldt country where competition was keener between swiftness and stealth, and where adroitness of thinking and movement played a preponderating role in the preservation of the species'.

He recalled how Darwin had predicted that Africa would prove to be the cradle of humankind. 'In my opinion', he wrote, 'Southern Africa, by providing a vast open country with occasional wooded belts and a relative scarcity of water, together with a fierce and bitter mammalian competition, furnished a laboratory such as was essential to this penultimate phase of human evolution'.

Even though Dart had not yet completed his work on digging out parts of the skull, he decided to despatch his findings to England. On 6 January 1925—only forty days after first catching sight of the Taung child—Dart posted his article to *Nature*, together with line drawings and photographs, in time to catch the Cape Town mail boat. He also alerted the local press. He anticipated a degree of scepticism from the British scientific establishment. What he did not expect was outright rejection.

Dart's manuscript arrived on the desk of the editor of *Nature*, Richard Gregory, on 30 January. Gregory considered its claims to be 'so unprecedented' that he sent proofs of the article to four eminent experts: Sir Arthur Keith, who was the current doyen of British evolutionary studies; Grafton Elliot Smith, the brain specialist at University College, London; Sir Arthur Smith Woodward from the Natural History Museum; and Wynfrid Duckworth, a Cambridge anatomist. But before the four experts had time to give it much consideration, news swept around the world that Dart had discovered 'the missing link'.

On 3 February, the Johannesburg *Star* published a scoop about the Taung child, based on Dart's article and photographs that he had given to its news editor. The *Star*'s report was carried by other newspapers the following day—Dart's thirty-second birthday—turning the fossil into a global sensation. Headlines focused upon some of Dart's more dramatic claims: 'Ape-Man of Africa had commonsense'; 'Missing Link that could speak'; 'Birth of Mankind'. For days, Dart was inundated with cables offering congratulations. Learned journals asked for articles; publishers proposed book contracts.

But the reaction of the scientific establishment was far more cautious. During the three years that Dart had spent in London, working at University College, he had gained a mixed reputation. He was seen as having high potential but also a troublesome streak. A paper he had produced on nerve cells challenging accepted opinion had turned out to be wrong; Dart's adamant defence of his position had raised concerns among some scientists that he 'might be inclined too hastily to arrive at conclusions on too little evidence'. Sir Arthur Keith recalled in his autobiography: 'Of his knowledge, his power of intellect and of imagination there could be no question; what rather frightened me was his flightiness, his scorn for accepted opinion, the unorthodoxy of his outlook'. Keith had been willing to recommend Dart for the Johannesburg post, but he had done so, he said, 'with a certain degree of trepidation'.

What immediately disturbed the scientific establishment was the speed with which Dart had leaped into print. The protocol they followed required scientists to spend months, even years, studying specimens before proffering their conclusions. Sir Arthur Smith Woodward had kept a skull from Singa in the Sudan for ten years before publishing a short report on it. The British Museum took seven years to publish a full assessment of a skull from Broken Hill in Northern Rhodesia (Zambia) found in 1921.

The scientific community also disliked Dart's use of extravagant speculation and florid prose.

But there were more profound reasons for the hostile reaction that Dart encountered. Prominent scientists such as Keith and Elliot Smith were convinced that the key factor enabling humans to emerge from the ape masses was brain power; a large brain, they insisted, had preceded the development of other faculties, such as upright walking. Keith had worked out a specific threshold needed for a specimen to be included in the genus *Homo*: a cranial capacity of 800 cubic centimetres or more. This theory about the importance of brain size had led Keith and Elliot Smith to validate the Piltdown skull as authentic. They continued to regard Piltdown Man as the most important discovery yet made in the search for human origins. By contrast, Dart's specimen had a small ape-sized brain.

Dart's article was published in *Nature* on 7 February, and one week later *Nature* published the comments of the four eminent experts it had solicited. All four emphasised the difficulty of assessing a fossil, especially a juvenile fossil, from a preliminary report and a few photographs.

But they all nevertheless detected more similarities with apes than humans.

Keith's opinion, in particular, carried enormous weight. He was a central figure in an international circle of distinguished scientists, holding high office in several scientific organisations. His initial response was guarded:

> It may be that *Australopithecus* does turn out to be 'intermediate between living anthropoids and man', but on the evidence now produced one is inclined to place *Australopithecus* in the same group or sub-family as the chimpanzee and gorilla. It is an allied genus. It seems to be near akin to both.

Elliot Smith was more sceptical. During Dart's tenure at University College, he had acted as his mentor and had been instrumental in persuading him to take up the post in Johannesburg. But now he was worried about the extent of Dart's claims. 'Many of the features cited by Professor Dart as evidence of human affinities, especially the features of the jaw and teeth mentioned by him, are not unknown in the young of the giant anthropoids and even in the adult'. And he asked for more proof. 'What above all we want Professor Dart to tell us is the geological evidence of age, the exact condition under which the fossil was found, and the exact form of the teeth'.

Smith Woodward was dismissive. 'I see nothing in the orbits, nasal bones, and canine teeth definitely nearer to the human condition than the corresponding parts of the skull of a modern young chimpanzee'. He challenged Dart's assertion about an African origin for humankind. 'The new fossil from Africa certainly has little bearing on the question'. And he concluded by regretting that Dart had chosen to use a 'barbarous' combination of Latin and Greek in naming the specimen *Australopithecus*.

Duckworth was more sympathetic. It was illustrations from Duckworth's treatise *Morphology and Anthropology* that Dart had used to make a comparison between the Taung specimen and apes. But he raised the question of whether the apparent humanlike features were not due to the young age of the specimen and concluded that the Taung child was most closely related to a gorilla.

In answering this barrage of criticism, Dart was severely hampered by the circumstances surrounding the discovery of the Taung child's skull. At the time the skull was picked up in the quarry, no attempt had been made to measure, photograph or accurately record the site's stratigraphy in a way that would have helped establish how old the specimen was. The all-important question of the age of the skull thus remained unresolved. Nor was Dart able to overcome the difficulty of proving that the 'humanoid' features to which he had pointed were not due to its childlike age; no adult specimen was available.

But the most formidable hurdle he faced was how to overcome objections about the size of its brain. Keith calculated that the infant *Australopithecus* possessed a brain capacity of less than 450 cubic centimetres; and that the brain capacity of an adult of its kind would reach no more than 520 cubic centimetres. It was thus hardly a suitable candidate for being in the direct line of human ancestors.

Worse was to follow. The experts had based their opinions entirely on Dart's description and illustrations. They were keen to see, if not the original specimen, then at least a cast (replica) of it. But Dart was slow to produce casts. No one in his department knew how to make them; nor did he. Eventually he hired a professional plasterer for the job.

But instead of giving the experts a preview of the casts, Dart sent them as exhibits to the South African pavilion at the British Empire Exhibition at Wembley in London that was due to open in June. There they were mounted in a glass cage under a banner proclaiming 'Africa: The Cradle of Humanity', with charts asserting the Taung child to be a direct human ancestor. To get a glimpse of the casts, the experts were obliged to gaze through the glass cage, while jostling with members of the public. Keith was furious. 'For some reason, which has not been made clear, students of fossil men have not been given an opportunity of purchasing these casts', he protested. 'If they wish to study them they must visit Wembley and peer at them in a glass cage'.

Outraged by Dart's conduct, the scientific establishment closed ranks against him. After observing the plaster casts, Keith rejected Dart's entire case. 'The famous Taung skull is not that of the missing link between ape and man', he said in a press statement. In a letter to *Nature* on 22 June, he reported: 'An examination of the casts exhibited at Wembley will satisfy zoologists that this claim is preposterous. The skull is that of a young anthropoid ape—one which is in its fourth year of growth, a child—and showing so many points of affinity with the two living African anthropoids, the gorilla and the chimpanzee, that there cannot be a moment's hesitation in placing the fossil form in this living group'. The Taung 'ape', he said, was 'much too late in the scale of time to have any place in man's ancestry'.

Elliot Smith followed suit. In a lecture at University College, he remarked: 'It is unfortunate that Dart had no access to skulls of infant chimpanzees, gorillas or orangs of an age corresponding to that of the Taung skull, for had such material been available he would have realised that the posture and poise of the head, the shape of the jaws, and many details of the nose, face and cranium upon which he relied for proof of his contention that *Australopithecus* was nearly akin to man, were essentially identical with the conditions in the infant gorilla and chimpanzee'.

Dart never recovered from these attacks. Not only scientific colleagues but popular opinion veered against him. The Taung child became little more than a music-hall joke. Disheartened by this turn of events, Dart buried himself in university work. When the university authorities offered him the opportunity to travel to Europe to show his prize specimen to scientists there and to compare it with other known fossils, he declined to go. Nor did he make any attempt to search for an adult *Australopithecus* at Taung or at other limestone mines to bolster his case.

Moreover, the flurry of interest in the Taung child was soon overtaken by news of a significant discovery in China. Palaeontologists working in an abandoned lime quarry at Chou K'ou Tien (now Zhoukoudian), a village forty miles from Beijing, uncovered hominid remains that became known as 'Peking Man', adding weight to the theory that Asia, not Africa, was the cradle of humankind. A distinguished American scientist, Henry Fairfield Osborn, director of the American Museum of Natural History, was convinced of the matter and steered large sums of money towards research in Asia. In a book published in 1927, *Man Rises to Parnassus*, Osborn made not a single reference to Dart, Taung or *Australopithecus africanus*. Not only was Dart's child the wrong creature; it was in the wrong part of the world.

When members of the British Association for the Advancement of Science visited South Africa in 1929, Dart's hopes that they would take an interest in the Taung child were soon dashed. 'Although some examined and made non-committal comments', he recalled, 'it was obvious that few regarded it as anything of real importance in the evolutionary story'.

Dart made one last attempt to persuade the scientific establishment of the validity of the Taung child. In 1931, six years after he had first set eyes on it, he brought the skull to London, hoping for a more favourable reception. But he found the London experts—Keith, Elliot Smith and Smith Woodward—far more preoccupied with Peking Man than interested in listening to his arguments. Elliot Smith, who had recently returned from China, was brimming with enthusiasm about the discoveries there. Dart was nevertheless invited to share a platform with Elliot Smith at a meeting of the Zoological Society of London in February.

With Smith Woodward in the chair, Elliot Smith led off with a masterly account of the Peking Man discoveries, enlivened by lantern slides and casts. Dart's heart sank: he had neither slides nor casts, only the tiny skull of *Australopithecus africanus* cradled in his hands.

> I stood in that austere and chilly room, my heart bounding with the hope that the expressions of polite attention on the four score faces before me might change to vivid interest as I spoke. I realized that my offering was an anti-climax . . .
>
> My address became increasingly diffident as I realized the inadequacy of my material and took in the unchanged expressions of my audience.

Further disappointment followed. Dart had arrived in London with high hopes that the Royal Society would publish a 300-page monograph he had written on the Taung child. But Elliot Smith informed him that only a section of it on dentition would be accepted. Rather than agree to such cuts, Dart took the monograph back with him to South Africa, abandoning plans to have it published.

The final blow came later in 1931, when Keith published his book *New Discoveries Relating to the Antiquity of Man,* in which he devoted an entire chapter to demolishing Dart's claims about the Taung child. In Britain, Keith's verdict was regarded as being the last word on the matter.

Demoralised and defeated, Dart lost all interest in palaeoanthropology, gave up work on fossils for many years and subsequently suffered a nervous breakdown. The Taung child, meanwhile, lay forgotten on the desk of one of his colleagues in the medical school.

There was one man, however, who took up the cause of Dart's child, with extraordinary results.

CHAPTER 3

Broom's Triumph

Robert Broom was both a man of genius and a rogue. He regarded himself as the greatest palaeontologist who had ever lived. His output of scientific papers was prolific. He was acknowledged to be a world authority on the mammal-like reptiles of the prehistoric Karoo, a semi-desert region of South Africa where he lived for many years. Yet for much of his career he had been treated as an outcast by scientific colleagues; at one stage he was banned from access to collections of the South African Museum that he himself had helped establish. His reputation for dubious practices frequently overshadowed his work. 'If one asks people who knew Broom well whether he was honest, the answers are a little confusing', wrote his biographer, George Findlay. 'He probably had the honesty of a good poker player'.

Born in Scotland in 1866, Broom trained as a medical doctor, graduating from Glasgow University in 1889, but was soon consumed by an interest in the origin of mammals. In 1892, he travelled to Australia, home of the most primitive of living mammals, working there as a doctor for four years but spending his spare time studying zoology. Papers he produced on the anatomy and embryology of Australian mammals marked him out as a promising young anatomist.

Returning to London in 1896, Broom became intrigued by fossils from the Karoo, held at the Natural History Museum, which appeared to have links with primitive mammals. He travelled to South Africa in 1897, found work as a medical locum in villages in Namaqualand and began collecting a wide variety of specimens, sending many of them to colleagues overseas. Among the specimens he sent to Sir William Turner at Edinburgh University were some human skulls taken from the bodies of Khoikhoi tribesmen who had died in a recent drought. 'I cut off their heads', he explained in a letter, 'and boiled them in paraffin tins on the kitchen stove'.

South Africa henceforth became his main home. In 1900, he set up a medical practice in the Karoo village of Pearston, devoting much of his time to hunting for fossils. In 1903, he was appointed Professor of Geology and Zoology at Victoria College in Stellenbosch; and from 1905, he also held the post of Curator of Fossil Vertebrates at the South African Museum in Cape Town. Given a free railway pass, he continued to explore the Karoo, assembling a collection of fossil finds that showed how a group of reptiles had gradually evolved into mammals. Over a period of seven years, he published more than 100 scientific papers.

But in 1909, he lost his free railway pass when a government minister decided that the study and collection of fossils was not a matter of national interest. He also fell foul of the museum authorities. In 1910, he decided to give up his college post and return to medical practice. He supplemented his income as a country doctor by running what amounted to a wholesale business in Karoo fossils, paying collectors to bring him specimens and selling them to clients abroad. His fall into notoriety came in 1913, when he sold a large and rare collection of Karoo fossils said to belong to the South African Museum to the American Museum of Natural History. 'Here I sit with a pocketful of dollars, and not a friend in the world', Broom wrote to a fellow collector.

In 1918, Broom settled in Douglas, a small town in the northern Cape on the edge of the Kalahari Desert, taking up an appointment as district surgeon but continuing his trade in fossils. He bought a large family house on Giddy Street, served on the municipal council and was elected mayor for five years.

The house on Giddy Street was soon reputed to be haunted. Broom was an avid collector of skulls and skeletons and used the house as a kind of laboratory. His son, Norman, recalled: 'It was not uncommon for a human skull or some other horror to be placed on the stove to cook merrily alongside whatever was being prepared for the next meal. Mother never took kindly to this and neither did the servants. Skulls would be left lying around in most of the rooms and it was never necessary to lock up the house for no strange native could be tempted to come near the place'.

Broom raided graves for research purposes and used other unconventional methods of obtaining bodies if he came across interesting specimens.

> If a prisoner dies and you want his skeleton [he recalled], probably two or three regulations stood in the way, but the enthusiast does not worry about such regulations. I used to get the body sent up . . . then the remains would be buried in my garden, and in a few months the bones would be collected.

He admitted that 'studying anthropology is not always a pleasant task'.

> One day a very interesting native died and I wanted the skeleton very badly so I had the body sent up to my garage for me to do a post mortem. It was in January and the temperature was much above 100 degrees in the shade. I was called out on a long country journey and only got back at 10 o'clock at night . . . I fear that the European armchair anthropologists have little ideas of the troubles we workers in the field have.

In his hunt for fossils Broom displayed remarkable stamina. Even under the hottest sun, he would invariably dress in a dark suit and waistcoat, long-sleeved white shirt, stiff butterfly collar and sombre tie, yet never show any signs of fatigue. Throughout his life he appeared to be in a hurry, walking and talking at a brisk pace, turning out scientific papers by the score. By 1925, he had written some 250 papers, named some seventy new genera and almost 200 new species of reptiles.

He also began to acquire an interest in early hominids after learning of the discovery of Boskop Man, the first fossil skull to be unearthed in South Africa. An incomplete fossil skull, it had been found in 1913 by two farmers digging an irrigation ditch at Boskop, near Potchefstroom, in western Transvaal. Broom examined Boskop Man in 1917 and wrote a paper claiming it to be a new species of primitive man that he called *Homo capensis*. His interest in early hominids deepened when he read reports of the discovery in Northern Rhodesia (Zambia) in 1921 of Rhodesian Man (*Homo rhodesiensis*), a human skull with beetling eyebrow ridges thicker than those of Neanderthal Man and with a muzzle as massive as a gorilla—but with teeth like a modern human and with a large brain.

Then in February 1925 came news of the discovery of the Taung child.

Broom's immediate reaction was to send Dart a letter of congratulations. Two weeks later, without any prior arrangement, he suddenly arrived at Dart's laboratory in Johannesburg. Although Dart knew of Broom's work, the two men had never previously met. To Dart's surprise, Broom walked straight past him and other members of his staff, strode over to the bench where the Taung fossil had been placed and dropped to his knees to examine it more closely. Looking up at Dart over his spectacles with a quizzical smile, he said: 'I am kneeling in adoration of our ancestor'.

Broom spent the weekend at Dart's home examining the skull carefully, becoming all the more certain of its rightful place in human ancestry. 'As a palaeontologist I did not greatly worry about the size and shape of the brain or the convolutions', he wrote, 'but I was convinced from the structure of the teeth that the [Taung] child was not allied to either the chimpanzee or the gorilla, and that it was closely allied to man'.

Having satisfied himself about Dart's claims, he sent an article to *Nature* in London and to *Natural History* in New York supporting him. 'We have a connecting link between the higher apes and one of the lowest human types', he wrote. And he told a *Cape Times* correspondent: 'The skull is probably the most important ancestral human skull found. In fact, I regard it as the most important fossil ever discovered'.

Like Dart, he was shocked by the reaction of the British scientific establishment. Writing his memoirs twenty-five years later, he was still incandescent:

> In England, many took little interest in the discovery of what might be a being closely related to man's ancestors, but they were greatly interested in the pedantic question of whether the name *Australopithecus* was good Latin! Prof. Dart might or might not be a great anatomist, but they were sure he was not a great classical scholar. As if it mattered in the least!

He recalled how a prominent scientist at the British Museum, F. A. Bather, had scolded Dart in the columns of *Nature*: 'If you want to join in a game, you must learn the rules', Bather had said. 'Professor Dart does not yet realize the many sidedness of his offences'. Broom fumed:

> It makes one rub one's eyes. Here was a man who had made one of the greatest discoveries in the world's history—a discovery that may

> yet rank in importance with Darwin's *Origin of Species*; and English culture treats him as if he had been a naughty schoolboy.

And he added caustically:

> I was never able to discover what were Prof. Dart's offences. Presumably the most serious was that when he found a very important skull he did not immediately send it off to the British Museum, where it would have been examined by an 'expert', and probably described ten years later, but boldly described it himself, and published an account within a few weeks of the discovery.

The outcome, Broom said, had been disastrous. 'Our wonderful South African "Missing Link" was discredited, and became a joke; and no one worried to look for more'.

Indeed, research work in South Africa came to a standstill for ten years.

Broom himself fell on hard times. During the Great Depression of the early 1930s, when the sheep-farming communities of the Karoo faced hardship and destitution, Broom too found it difficult to make ends meet. In 1933, he was elected president of the South African Association for the Advancement of Science but was unable to afford the train fare to attend its annual conference and needed help. When Dart realised his circumstances, he appealed to General Jan Smuts, a senior government minister with a keen interest in natural history and human evolution, to rescue Broom, pointing to the waste of talent. In 1934, Broom was duly appointed Keeper of Vertebrate Palaeontology and Anthropology at the Transvaal Museum in Pretoria, though not without misgivings on the part of the museum authorities. He also joined Dart's anatomy department as a lecturer in comparative anatomy. Over the next two years, he worked on the museum's collection of fossil reptiles, writing sixteen papers for publication, identifying twenty-three new genera and forty-four new species.

Then, at the age of sixty-eight, he decided to embark on a new career. Having become the greatest palaeontologist who had ever lived, he remarked, he saw no reason he should not become the greatest palaeoanthropologist as well.

Broom's aim was to find an adult Taung 'ape' to prove that australopithecines were in the direct line of human ancestors. Two of Dart's students drew his attention to a huge underground cave at Sterkfontein, forty miles west of Johannesburg. The cave had been discovered in the 1890s during blasting operations at a limestone quarry. A visitor in 1898 recorded that its entrance was as grand as 'the hall of a mansion'. Inside, he wrote, 'Thousands of stalactites of different shapes and sizes hang above one's head. Several have long since become joined to stalagmites and form magnificent pillars'. Later in the 1920s, with public interest in such sites stimulated by the discovery of the Taung child in 1924, the Sterkfontein cave had attracted a growing number of tourists and souvenir hunters. An enterprising local store-owner who traded in bat guano from the caves had written a brief guide to the site, urging readers to 'Come to Sterkfontein and buy your guano, and find the missing link'. But no serious scientific work had ever been carried out there.

Accompanied by Dart's students, Broom arrived at Sterkfontein on 9 August 1936, dressed, as ever, in a dark three-piece suit and wing collar. By coincidence, the quarry supervisor at Sterkfontein, George Barlow, had been the manager at Taung when the first australopithecine had been found. He still took an interest in fossils, selling them from a tea-room to visitors. Broom asked him whether he had seen anything at Sterkfontein like the Taung skull, and Barlow replied that he thought that he had. Broom asked him to keep a lookout for any promising specimens.

Three days later, when Broom returned, Barlow handed him three baboon skulls and part of a sabre-toothed cat skull. When he returned again on 17 August, Barlow showed him a far more significant find: a fossilised brain-cast. It had been blasted out that morning. 'Is this what you're after?' asked Barlow.

Broom scoured the blasted area for other parts, but with no success. The following day, however, he returned with a museum team and found the base of the skull and a number of bone fragments. After several weeks of work, he managed to assemble a rather battered and incomplete skull of what seemed to be an *Australopithecus*. He placed the find in a new species, calling it *Australopithecus transvaalensis*, but later decided, because of differences he perceived between the teeth of the Taung and the Sterkfontein specimens, to move it into a new genus, calling it *Plesianthropus transvaalensis*—'near man of the Transvaal'.

Like Dart, he lost no time in publicising his find, sending accounts to *Nature* and to *Illustrated London News*, insisting it confirmed Dart's views. The *Illustrated London News* ran a summary of his article in September under the heading: 'A new Ancestral Link between Ape and Man'. But the scientific establishment still preferred to regard Asia rather than Africa as the birthplace of mankind.

Two years later, Broom made another breakthrough. In June 1938, Barlow handed him what appeared to be an australopithecine palate with one molar still in place. It had been given to Barlow by a fifteen-year-old schoolboy, Gert Terblanche, who had found it on a hillside on a neighbouring farm called Kromdraai, a mile from the Sterkfontein site. Broom immediately set out to investigate and tracked down Gert at his school. Summoned by the headmaster, Gert produced from his trouser pocket what Broom described as 'four of the most wonderful teeth ever seen in the world's history'. Gert explained that he had prised the teeth from a jawbone embedded in rocks on the hillside. Broom purchased them on the spot for a shilling apiece and, after enthralling pupils and teachers with an impromptu lecture on fossils and cave formations, headed with Gert to the hillside on Kromdraai, where Gert retrieved the jawbone from a hiding place. Although much of the Kromdraai skull had been smashed, Broom arranged for every fragment of bone and tooth to be collected.

When the skull was reconstructed, the Kromdraai fossil turned out to be different from the Sterkfontein specimen. Its face was flatter, its jaw was more powerful and its teeth were larger. Broom therefore decided to allocate it to yet another genus and species, calling it *Paranthropus robustus*—'robust creature next to man'.

Once again, he was swift to send accounts to *Nature* and to *Illustrated London News*. The *Illustrated London News* acclaimed the new find under the headline: 'The Missing Link No Longer Missing'. The scientific establishment in Britain, however, remained sceptical. He was criticised for creating new genera on 'extremely slender grounds' and told to act with greater caution. 'The English are not accustomed to such daring', observed Broom.

But scientific opinion in the United States was beginning to turn. In June 1938, two influential scientists, William King Gregory and Milo Hellman of the American Museum of Natural History, arrived in South Africa to examine the original specimens from Taung and Sterkfontein. Gregory had previously dismissed the Taung child as no more than an ape. The two Americans now concluded that the specimens were 'in both a structural and a genetic sense the conservative cousins of the contemporary human branch'. At a meeting of the Associated Scientific and Technical Societies of South Africa in July 1938, they paid tribute to Dart and Broom. 'The whole world is indebted to these two men for their discoveries, which have reached the climax of more than a century of research on that great problem, the origin and physical structure of man'. The following year they placed all three specimens—*Australopithecus*, *Plesianthropus* and *Paranthropus*—within the same subfamily, Australopithecinae, of the family Hominidae. But many other American scientists remained sceptical.

During the war years, Broom compiled all the evidence about the fossils found at Taung, Sterkfontein and Kromdraai in a comprehensive volume entitled *The South African Fossil Ape-Men: The Australopithecinae*. It was published in 1946, shortly after his eightieth birthday. Broom concluded that australopithecines resembled humans in several ways. 'They were almost certainly bipedal and they probably used their hands for the manipulation of implements'. Although they had small brains and apelike faces, their teeth were similar. 'What appears certain is that the group, if not quite worthy of being called men, were nearly men, and were closely allied to mankind, and not at all nearly related to the living anthropoids'. Altogether, wrote Broom, 'if one could be found alive today I think it probable that most scientists would regard him as a primitive form of man'.

Broom was given an award by the U.S. National Academy of Sciences for producing the year's most important biological book. Although critics in England accused Broom of being too ambitious and too hasty in reaching his conclusions, even there the tide of opinion began to turn. At the end of 1946, an influential Oxford anatomist, Wilfrid Le Gros Clark, spent two weeks in South Africa poring over the fossils and visiting cave sites. He arrived, he recalled, as the 'devil's advocate', bent on opposing Broom's claims, but was soon convinced that he was right.

When scientists gathered for the first Pan-African Congress on Prehistory in Nairobi, the capital of Kenya, in January 1947, the South African fossil finds were consequently the centre of attention. Dart and Broom were invited to give a presentation of their work. Le Gros Clark followed, throwing his weight behind them. 'I am afraid there is no escape from the fact that these specimens are very closely related to man and are survivors of the group that gave origin to man', he told the assembled scientists. The australopithecines, he said, were 'man in the making'. Le Gros Clark's verdict stunned his colleagues and made an impact around the world. 'The suggestion that the Australopithecines are to be regarded as anthropoid apes . . . must almost certainly be ruled out', a correspondent in *Nature* reported from the conference. 'There appeared no room for doubt that Dart and Broom had certainly not over-estimated the significance of the Australopithecinae, and their interpretations of these fossil remains were entirely correct in all essential details'.

Relishing the attention, Broom impressed all with his boundless enthusiasm. During an excursion to a rock art site near Kisese, he ignored a plea from the Kenyan archaeologist Louis Leakey to forgo the two-mile walk to the cave paintings in the blazing summer sun. Leakey wrote: 'I shall never forget the sight of Robert Broom . . . wearing, as always, a dark suit, wing collar and butterfly tie, negotiating the last steep stretch in the heat of the day. It was indeed an amazing feat for a man of his age in such unsuitable clothing'. When the visitors were obliged to cross a river in flood on foot, it was Broom who led the way, with his black trousers rolled up to his knees.

The 1947 Congress was not only a personal triumph for Dart and Broom. It marked the point at which scientific opinion began to consider Africa rather than Asia as the more likely birthplace of humankind. Buoyed up by their discussions, delegates resolved to hold conferences on African prehistory on a regular basis every four years, accepting an invitation from South Africa's prime minister, Jan Smuts, to meet in South Africa in 1951.

On his return to South Africa, Broom threw himself with gusto into the hunt for more fossils, helped by a talented young assistant, John Robinson. Smuts promised to provide him with government funds. But Broom was soon embroiled in further controversy. His colleagues became increasingly concerned about his slapdash methods of recording his work. A Cambridge-trained archaeologist, Basil Cooke, recalled how in 1947 he and a colleague visiting the Transvaal Museum asked

Broom if they could see the Kromdraai skull. 'Broom fossicked in a drawer and pulled out the facial part, then rushed off down the corridor, with us on his trail, and into the laboratory of Vivien Fitzsimmons. There he pushed aside two jars of snakes and said: "Here's the other piece. I knew it must be there"'.

There was also alarm about Broom's methods in the field, in particular his liberal use of dynamite to blast fossils from rock-hard breccia. Geologists complained that indiscriminate dynamiting destroyed the stratigraphic context of the fossils, making it difficult if not impossible to try to date them. To Broom's fury, the Historical Monuments Commission intervened, insisting that Broom would not be allowed to continue his work unless he was assisted by a 'competent field geologist'.

Always ready for a fight, Broom refused to comply. 'I regarded it as an insult', he recalled. 'I had no compunction whatever about breaking the law. I considered that a bad law ought to be deliberately broken'. He appealed to Smuts for support and carried on blasting at Sterkfontein.

He was soon vindicated. On 18 April 1947, as the smoke from blasting drifted away, Broom recovered what he described as 'the most important fossil skull ever found in the world's history'. The blast had split the skull into two fragments but had not damaged it irreparably. It was complete but for the teeth and the lower jaw. 'I have seen so many interesting sights in my long life', Broom recalled, 'but this was the most thrilling in my experience'.

The skull was clearly that of an adult australopithecine—an adult version of the Taung child. After thorough inspection, Broom believed it to have belonged to a middle-aged female. He labelled it *Plesianthropus africanus*. But it became more popularly known as 'Mrs Ples'. It proved Dart's case, twenty-two years after he made it, that the Taung child was not simply a juvenile ape.

While the discovery of Mrs Ples was acclaimed in Europe and the United States, members of the Historical Monuments Commission were enraged that Broom had so brazenly defied their ruling about the use of dynamite and banned him from the Sterkfontein site. But after a period of public ridicule, they were obliged to relent.

Broom's spectacular run of luck at Sterkfontein continued throughout 1947. In August, he blasted out a chunk of breccia containing two sides of a pelvis, mostly intact, providing crucial evidence that australopithecines had been able to walk upright. Together with John Robinson, he opened up a new excavation site at Swartkrans, a disused lime quarry across the valley from Sterkfontein, making further discoveries, including a nearly complete mandible that he named *Telanthropus capensis*.

Even diehard critics were impressed. In a letter to *Nature* in 1947, Sir Arthur Keith conceded: 'I am now convinced on the evidence submitted by Dr. Robert Broom that Professor Dart was right and I was wrong'. The following year, in a book entitled *A New Theory of Human Evolution*, Keith agreed that 'of all the fossil forms known to us, the Australopithecinae are the nearest akin to man and the most likely to stand in direct line of man's ascent'.

In his final years, Broom endeavoured to make clearer sense of the mass of evidence he had accumulated. Despite his penchant for pinning different names to his collection of fossils, he nevertheless assigned them to a single subfamily, the Australopithecinae. Two main species had emerged: the lightly built 'gracile' *Australopithecus africanus* (which included the Taung child and Mrs Ples); and the 'robust' australopithecines like *Paranthropus* (later renamed *Australopithecus robustus*). Both walked upright; both possessed relatively small brains; both had teeth that were humanlike. Although the age of cave fossils was difficult to determine, it seemed likely that *africanus* predated and was ancestral to *robustus*. It also seemed likely that *africanus* was the ancestor of the line that led to *Homo sapiens*.

By the time he died on 6 April 1951, at the age of eighty-four, Broom had transformed the study of palaeoanthropology. Australopithecines, henceforth, became a recognised landmark on the path of human evolution. Moreover, by identifying two types of australopithecines of similar appearance, Broom had opened up an entirely new prospect.

'Since one of the two ape-men seemed clearly to be on the line of human descent and the other to have specialised away from that line', wrote Phillip Tobias, a distinguished South African scientist from the next generation, 'Broom's finds compelled scholars to realise that not all early hominids were direct ancestors of modern mankind. Some were on side branches. This meant that at an earlier period the two species so closely related to each other must have branched off from a common ancestor. The pattern of hominid evolution was not like a linear Chain of Being after all. It was like a bush of branches, only one of which made the grade to the later stages of human evolution, while other branches were doomed to ultimate extinction'.

But just when South Africa seemed to be leading the world in the search for human origins, a dark age settled over the country. In 1948, Afrikaner Nationalists gained power, determined to enforce white supremacy under a system of apartheid and hostile to any notion that whites and blacks might have shared a common humanity. The new National Party government held that blacks were an inferior race, destined by the will of God to be hewers of wood and drawers of water, and quoted texts from the Bible as proof. Nationalist politicians scorned the notion that humankind could have descended from an ancient African ape and promoted the teaching of creationist views. State schools were barred from teaching evolutionary theory. When the government gave notice that it would not allow black delegates to attend the second Pan African Congress on Prehistory, scheduled to be held in South Africa in 1951, international support dwindled, too; the congress had to be relocated to Algeria.

The focus of attention switched instead to East Africa.

CHAPTER 4

WHITE AFRICAN

LIKE RAYMOND DART and Robert Broom, Louis Leakey was regarded by the scientific establishment as a maverick, best kept at a safe distance. His early career as a field scientist had been full of promise, but he had since become immersed in controversy and scandal. No one doubted his energy, enthusiasm and talent, but colleagues found him too impetuous, too dogmatic, too intolerant of criticism, too much of a showman.

The son of an English missionary, born in 1903 on a mission station at Kabete in the hills above Nairobi, Leakey had grown up among the Kikuyu people, imbibing Kikuyu customs and folklore. At the age of thirteen, he had been initiated as a member of the Mukanda age group. 'In language and in mental outlook I was more Kikuyu than English', he wrote in his book *White African*, 'and it never occurred to me to act other than as a Kikuyu'. Chief Koinange of the Kikuyu spoke of him as 'the blackman with a white face', accepting him as 'one of ourselves'.

Leakey also acquired an early fascination with stone tools. His interest was prompted by a book on the 'Stone Age Men' of Britain, sent to him by a cousin in England as a Christmas present when he was twelve. Reading about how they had used flint arrowheads and axe heads, he set out to see if he could find some around Kabete. He had no clear idea what flints looked like, only that they were blackish in colour. Exploring road cuttings and other areas of exposed ground, he soon collected a mass of black flakes of rock that seemed to correspond to the flint tools he had read about. But his parents were sceptical and referred to them as 'broken bottles'. His Kikuyu friends were also doubtful. Noticing how black flakes appeared on the ground after heavy bouts of rain, their explanation was that they had fallen from the sky; they called them *nyenji cia ngoma*—razors of the spirits of the sky.

Leakey eventually resolved to show his collection to Arthur Loveridge, the curator of the Nairobi Museum. He arrived fearing that Loveridge too might laugh at him but was 'delighted beyond words' by his response. Loveridge explained that his stones were not flint but obsidian—a black volcanic glass that produced a sharp cutting edge when flaked—and that obsidian was known to have been used for toolmaking in the past. Indeed, said Loveridge, several of the specimens in Leakey's collection had undoubtedly been fashioned for tool use. From that moment, Leakey became addicted to prehistory.

Sent to school in England at the age of sixteen, Leakey endured two miserable years there, making few friends, but managed to gain a place at Cambridge to study anthropology. He impressed fellow students at Cambridge with his energy, enthusiasm and passion for prehistory but was also noted for being brash and impetuous—'overcharged and unbalanced and unlikely to make good', according to one contemporary.

During his university years he gained valuable field experience spending eight months in southeast Tanganyika with a British Museum expedition hunting for fossils at Tendaguru, the site where Hans Reck's team had discovered the remains of a *Braciosaurus* in 1912. He also excelled in studies of anthropology and archaeology, becoming all the more convinced that Africa was the place to search for the origins of humankind, not Asia.

Shortly after graduating with a 'double first' in 1926, he set out on what he grandly called the East African Archaeological Expedition, to look for human fossils in Kenya. A Cambridge professor tried to dissuade him: 'Don't waste your time. There's nothing of significance to be found there. If you really want to spend your life studying early man, do it in Asia'. But Leakey was adamant. 'No', he replied. 'I was born in East Africa, and I've already found traces of early man there. Furthermore, I'm convinced that Africa, not Asia, is the cradle of mankind'.

The East African Archaeological Expedition, financed by a variety of grants, consisted of Leakey and one assistant. For a year Leakey explored caves and burial sites among the lakes and volcanoes of Kenya's Great Rift Valley, collecting a mass of stones and bones. His account of the expedition—'Stone Age Man in Kenya Colony'—appeared in *Nature* in July 1926, earning him recognition within the scientific community.

Back in England, Leakey fell under the spell of Sir Arthur Keith, spending hours working on fossil material at his laboratories at the Royal College of Surgeons. As Keith's disciple, he became an ardent advocate of the 'big-brain' theory of human development, arguing that because the human brain could only have developed to such a size over a prolonged period, the separation of humans from apes must have occurred far back in antiquity, as far back as the beginning of the Miocene period, then dated at about 1 million years ago. Leakey also supported Keith's contentions about the validity of Piltdown Man.

After raising funds for a second, larger expedition, Leakey returned to Kenya in 1928, accompanied by his newly married wife, Frida Avern, focusing again on Rift Valley sites. Throwing himself tirelessly into the work, he made significant discoveries of stone tools, including ancient hand-axes, and managed to piece together for the first time a sequence of prehistoric cultures in Kenya.

Yet his Cambridge mentors sometimes fretted about his propensity for grandstanding and overstatement. On learning that Leakey intended to attend a meeting of the British Association for the Advancement of Science in Johannesburg in 1929, Alfred Haddon warned him in a letter: 'Do not go in for wild hypotheses. These won't do your work any good and it's foolish to try to make a splash'. A distinguished East African geologist, E. J. Wayland, cautioned him against making 'over-emphatic' comments. 'Believe me you will serve archaeology, the expedition and yourself best by maintaining a strictly scientific attitude', he wrote in a letter. 'You have the chance of making yourself, in time, one of the leaders of archaeological thought—don't spoil your chances, for by doing so you will unintentionally let the science down'.

On this occasion, Leakey heeded their advice, delivering a restrained account of his work. To his delight, he found himself the centre of attention, while Raymond Dart and his Taung fossil stirred little interest.

No sooner had Leakey returned to England to widespread acclaim than he began to plan for his third expedition. His main objective this time was to solve the mystery of Oldoway Man, the skeleton that Hans Reck had discovered in Olduvai Gorge in 1913. Leakey had first studied Oldoway Man during a visit to Munich in 1927, and he went back there in 1929 to examine it further. The conclusion he reached was that it was not as ancient as Reck claimed but was of similar age to some of the human skeletons he had found during his expeditions in Kenya—Late Stone Age specimens dating back no more than 20,000 years.

Leakey discussed the matter with Reck in Berlin and invited him to join the expedition. Reck eagerly accepted, but he remained adamant about the ancient origins of Oldoway Man. Examining Reck's collection of Olduvai rocks and fossils, Leakey noticed similarities to stone tools he had picked up in Kenya and suggested that they might find other examples at Olduvai. But Reck disagreed. He had searched the area diligently for stone tools over a period of three months, he said, and found none. But Leakey persisted. 'I . . . made a small bet that I would find Stone Age implements at Oldoway within 24 hours of arriving there'. The wager was for £10—'a not inconsiderable part of my research funds for that year'.

Leakey's third expedition set out from Nairobi in September 1931. The 260-mile journey to Olduvai, along rough tracks and across unmapped stretches of the Serengeti Plains, took four days. As the convoy of vehicles approached the gorge, Leakey and Reck took the lead. 'Reck could hardly hide the emotion he was feeling at once more returning to the scene of his very great scientific discoveries', wrote Leakey. At first light, the next morning, while others slept, Leakey set off in search of stone tools. From his Kenya experience, he knew that stone tools were likely to be made not from flint, as Reck had tried to find, but from volcanic lava, chert or quartz. He soon found a perfect specimen of a hand-axe. 'I was nearly mad with delight and I rushed back with it into camp'. Within four days, seventy-seven more hand-axes were collected.

Exhilarated by the find, Leakey went on to inspect the site in Bed II, where Reck had discovered Oldoway Man. By good fortune, four wooden pegs that Reck had used to mark the site were still in place. Reck recounted how he had found the skeleton, insisting that it was not merely a recent burial, as Leakey and others had argued, but was as old as Bed II itself. Leakey was soon persuaded that Reck was right. In high spirits, Leakey, Reck and Arthur Hopwood, a British Museum palaeontologist, sat down to compile a report to *Nature* supporting Reck's original conclusion about the age of the skeleton. One week after arriving at Olduvai, Leakey was on his way back to Nairobi, convinced that his expedition had solved the mystery of Oldoway Man. In a short article he wrote for the London *Times*, before returning to Olduvai, he claimed that it was 'almost beyond question' that Oldoway Man was 'the oldest known authentic skeleton of *Homo sapiens*'.

For two months, the run of discoveries continued. Leakey described Olduvai as 'a veritable paradise for the prehistorian as well as for the palaeontologist'. His team recovered hand-axes from all five beds in the gorge, providing him with 'a complete sequence of evolutionary stages of the hand-axe culture'. In the oldest bed—Bed I—they found 'pebble tools', simple flakes struck off pebbles, which came from the earliest known culture in the world, named by Leakey as the Oldowan.

But the harsh conditions at Olduvai were a constant problem. Water was in short supply and had to be rationed. A strong wind blew incessantly, carrying with it swirls of fine black dust. Leakey wrote:

> If you spread some semi-liquid sun-melted butter on a piece of bread it would be covered with fine black dust before you could get it to your mouth. If you poured out a cup of tea or coffee in a few minutes it had a fine black scum of dust on its surface. You breathed dust-laden air, your nostrils were filled with dust, you ate dust, drank dust, slept in dust-ridden bedding, and in fact everything was dust, dust, dust! The heat of the sun was terrific; and if you had a tendency to perspire at all you did so very freely, and the dust mingled with the sweat to make your body filthy. And yet water was so scarce . . .

The team also had to contend with marauding lions, rhinoceroses and hyenas that frequented the gorge. By the end of two months they were 'really rather glad at the prospect of a change'.

After sorting out his collection in Nairobi, Leakey set out on another expedition, this time to fossil sites at Kanjera in western Kenya, eight miles from Lake Victoria. The conditions were different, but just as arduous. The area was frequently drenched by heavy downpours. 'Whereas the constant trouble at Oldoway had been "not enough water"', wrote Leakey, 'here our trouble was too much of it'. Hordes of mosquitoes swarmed around the camp site at night.

> Sometimes they were so bad after dark that it was really difficult to feed ourselves at supper time. We used to wear long trousers tucked into Wellington boots to protect our legs, and tie towels round our heads leaving only the mouth, nose and eyes exposed, but even so we were terribly bitten, and on one occasion one of us killed over a hundred mosquitoes on his face during one meal.

Within a matter of weeks, Leakey and his team made two significant discoveries: fragments of two skulls from Kanjera and part of a jaw from a site at Kanam, three miles away. The Kanam mandible was found in a block of rock dug out of the side of a gully on 29 March by one of Leakey's African assistants, Juma Gitau, who earlier that day had recovered the tooth of a *Deinotherium*, an extinct type of elephant, from an adjacent spot in the same cliff.

Leakey was confident that the mandible was extremely old, dating back to early Pleistocene times, more than 500,000 years. He lost no time in alerting the outside world to his discovery, sending a despatch to *Nature* on 19 April.

'The importance of this Kanam mandible', Leakey wrote in a subsequent book, *The Stone Age Races of Kenya*, 'lies in the fact that it can be dated geologically, palaeontologically and archaeologically, and that it represents the oldest known human fragment yet found in the African continent . . . It is not only the oldest known human fragment from

Africa, but the most ancient fragment of true *Homo* yet discovered anywhere in the world'.

Leakey earned high praise for his expeditions in East Africa. At a meeting in Cambridge in March 1933, when a group of twenty-six scientists gathered to review his work, Leakey was congratulated for the 'exceptional significance' of his discoveries. The London *Times*, reporting on the outcome of the meeting, suggested that Leakey's work had lent plausibility to the theory that 'Africa is the cradle of the human race'.

But his downfall soon followed.

The first dent to his reputation came from the findings of independent geologists who made a series of tests on Oldoway Man and the surrounding soil samples where it had been found. Their conclusion was that the body had been buried in Bed II in comparatively recent times. (Subsequent Carbon-14 tests dated the skeleton to 19,000 years ago.)

Despite the evidence, Leakey fought on for months in defence of his views. 'He made a bit of a fool of himself by his vehement insistence', recalled John Solomon, a geologist colleague. 'It showed that his attitude in those years was not that of a "scientist", but of an "enthusiast"'.

A far more damaging controversy erupted over the Kanam mandible. In 1934, an eminent geologist, Percy Boswell, Professor of Geology at Imperial College, London, arrived at the sites at Kanjera and Kanam to inspect Leakey's fieldwork. A stickler for detail, Boswell had been sceptical from the outset about Leakey's claims. He had previously played a leading role in demolishing Leakey's arguments about the age of Oldoway Man. To allay his concerns, Leakey invited him to visit the Kenya sites while he was there on his fourth expedition.

For Leakey, the trip proved to be a disaster. He found difficulty in identifying the exact location of the discoveries made three years before. Not only had he failed to make proper geological maps at the time, but iron pegs that he had cemented into the ground to mark the spot had meanwhile been removed by local fishermen to make fishing harpoons and spears. The landscape, moreover, had been altered by erosion from heavy rainfall.

Even worse, a photograph that Leakey had used to illustrate the position of the Kanam mandible turned out to record not the mandible site itself but another location several hundred yards distant. Leakey's own camera had malfunctioned in 1932, so afterward he had borrowed a friend's photograph instead. When the friend remarked: 'I am not sure that this is the exact spot', Leakey is said to have replied, 'Near enough'. The photograph had been displayed at an exhibition at the Royal College of Surgeons, alongside the jaw fragment. It was due to be published on the opening page of Leakey's forthcoming book, *The Stone Age Races of Kenya*. In haste, Leakey was obliged to cable Oxford University Press asking the publisher to hold distribution of the book until an erratum slip had been inserted.

Boswell was distinctly unimpressed. 'The Professor is in a bad humour over it', Leakey recorded in his diary. On his return to England, Boswell sent a scathing account to *Nature*, published in March 1935, not only accusing Leakey of incompetence but implying he had fabricated evidence. 'It is regrettable that the records are not more precise', Boswell concluded, 'and it is disappointing after the failure to establish any considerable geological age for Oldoway Man . . . that uncertain conditions of discovery should also force me to place Kanam and Kanjera man in a "suspense account"'.

Newspapers around the world picked up the story, reporting that Leakey's claim to have found 'the Oldest Fragment of Man' had been debunked.

Leakey returned to England in September 1935, his reputation severely damaged. He incurred further opprobrium after forsaking his wife, Frida, and their two children to live 'in sin' with a talented young illustrator, Mary Nicol. Divorce proceedings added to his notoriety. His Cambridge college terminated his research fellowship and the university authorities made clear they were not willing to consider him for an academic post. In dire financial straits, Leakey accepted an offer from the Rhodes Trust to undertake a detailed study of the Kikuyu people.

In January 1937, he set sail for Kenya, accompanied by his newly married wife, Mary, with little prospect of being able to pursue his search for the earliest man.

CHAPTER 5

Dear Boy

When Mary Nicol first arrived in East Africa in 1935 to join an expedition to Olduvai that Leakey had organised, she was given a swift introduction to the hazards of life with Louis in Africa. The 'long rains' had begun, turning roads into quagmires. Leakey was a day late in collecting Mary at Moshi, a small town at the foot of Mount Kilimanjaro, to which she had flown to meet him; on one stretch of the road from Nairobi, it had taken him five hours to cover 200 yards. They started out from Moshi well enough, hoping to reach the rim of Ngorongoro Crater in little more than a day, but it took them two days just to reach the bottom of the escarpment. With food supplies running low, they spent the night in a damp hut abandoned by road-makers. It rained all night long. The next day, slithering through the mud up the escarpment, they covered only three miles. 'On some occasions, [we] practically carried the car and the equipment', wrote Leakey. 'Sometimes we unloaded the car and carried the luggage ahead for half a mile or so, and then returned to push and carry the car. Sometimes we took the car ahead and went back to get the luggage'. Six days after they set out from Moshi, they reached the rim of the crater, black with mud from head to foot, tired and very hungry.

The daughter of an itinerant English landscape painter, Mary Nicol, born in 1913, had been educated at a succession of schools in France and England, gaining a reputation for being a troublemaker. She was expelled from her final school, accused of simulating a fit and causing an explosion during a chemistry lesson. She left without ever having passed a single school examination, but discovered a liking for archaeology, attended lectures at London's University College and developed an exceptional talent for drawing. A distinguished archaeologist, Dr Gertrude Caton-Thompson, was so impressed by Mary's work that she commissioned her to illustrate stone tools for her forthcoming book on the Fayoum depression in Egypt, *The Desert Fayoum*. She also introduced her to Louis Leakey, thinking he too might be in need of an artist to illustrate his book on early man, *Adam's Ancestors*. When Leakey saw Mary's drawings for *The Desert Fayoum*, he was struck with admiration. 'Mary Nicol's drawings', he recalled years later, 'were the best representations of stone tools I had ever seen then or, indeed, have seen since'. He immediately invited her to undertake the drawings for his book. By the summer of 1933, they were 'inseparable companions'. Mary was just twenty years old; Leakey was nearly thirty.

Mary took to the spartan life at Olduvai with relish. The surrounding plains were alive with great herds of wildebeest and zebra migrating after the rains; at night lions prowled around the edges of the camp. The gorge itself offered endless possibilities for exploration. 'In 1935 it was just a place that was incredibly beautiful and where nearly every exposure produced some archaeological or geological excitement', she recalled in her memoirs. Each day, Leakey's team headed out to explore new terrain, returning tired and dirty but fulfilled. Through diligent work, Mary slowly won the approval of other members of the expedition initially sceptical about the reasons for her presence. One of the gullies in the gorge was named after her. But camp conditions became increasingly arduous. When the rains stopped, water supplies became scarce. Low on fuel, Leakey could not spare enough petrol for vehicles to fetch fresh water from springs eighteen miles away and had to rely on dwindling pools in the gorge.

'Our water hole near the camp was little more than a liquid, muddy swamp, in which a rhino wallowed daily and, after wallowing, added a certain amount of urine', wrote Leakey. 'Not unnaturally, the water tasted very unpleasant and was not really fit to drink, but it was all we had'. Leakey's attempts to devise a filter system of charcoal and sand made little difference. 'Our soup, tea or coffee all tasted of rhino urine', Mary recalled, 'which we never quite got used to'. For all the hardships, however, she thrived on the adventure. When the expedition was over, she left for England knowing, she wrote, that she would never be the same again now that Africa had 'cast its spell' on her.

Returning to Kenya as a married couple in 1937, the Leakeys endeavoured to keep up their work on archaeological sites. Accepted as a professional archaeologist in her own right, Mary started excavations at a neolithic burial site at Hyrax Hill, a rocky ridge overlooking Lake Nakuru. Her work at Hyrax Hill was subsequently to gain high praise. She pioneered a technique of leaving stones and bones in exactly the same position in which they had been found, enabling other archaeologists to study their geological context, an innovation that set a new standard for excavation.

During the Christmas holiday in 1937, the Leakeys began digging at a burial site at Njoro on a farm belonging to Nellie Grant, mother of the writer Elspeth Huxley. But they were always short of funds, as Nellie Grant remarked in a letter:

> The poor Leakeys are held up for wages money and only have two months' work in sight on the skeletons, and need two years. They

live on the smell of an oil-rag themselves, work all day on the site and up to eight or nine o'clock at night, sorting and labelling the day's finds.

While the Leakeys were struggling to survive in Kenya, a new revolution in scientific thinking about evolution was gathering momentum in the West. For several decades, the field of evolutionary studies had been convulsed by competing theories. Some biologists held fast to the concept of orthogenesis, the idea that evolution was directed by an inner drive towards a fixed goal; some favoured transmutationism, the theory that evolutionary change occurred through sudden mutations, or saltations, producing new species instantaneously; others subscribed to the idea of some sort of inheritance of acquired characteristics. Darwin's theory of natural selection had long since fallen out of favour.

The new science of genetics added to the mix. Since 1900, geneticists had made great strides in working out the basic principles of inheritance, first enunciated by the Austrian monk Gregor Mendel in an obscure publication in 1865 but left unnoticed for thirty-five years. Mendel had discovered that physical traits were determined by stable inheritance factors, subsequently known as genes. By breeding strains of pea plants in his monastery garden, he found he could produce simple and mathematically predictable inheritance patterns. Inheritance, he showed, did not blend the genes of both parents; they remained discrete, so that rare genes from one parent could seem to vanish for a generation but reappear fully functional in the next generation. In the 1900s, a Dutch botanist, Hugo de Vries, added the theory that genes were capable of spontaneous changes, or 'mutations' that opened the way for further genetic variation. He argued that it was 'mutation pressure'—the rate at which mutations occurred—that was the driving force behind evolution. This theory of genetic mutation became a popular explanation for evolution in the early decades of the twentieth century.

But different schools of evolutionists—experimental geneticists, naturalists, palaeontologists and developmental biologists—tended to regard each other with mutual incomprehension and pursued their own research separately. As the American palaeontologist George Gaylard Simpson noted in 1944:

> Not long ago, paleontologists felt that a geneticist was a person who shut himself in a room, pulled down the shades, watched small flies disporting themselves in milk bottles, and thought that he was studying nature. A pursuit so removed from the realities of life, they said, had no significance for the true biologist. On the other hand, the geneticists said that paleontology had no further contributions to make to biology, that its only point had been the completed demonstration of the truth of evolution, and that it was a subject too purely descriptive to merit the name 'science.' The palaeontologist, they believed, is like a man who undertakes to study the principles of the internal combustion engine by standing on a street corner and watching the motor cars whiz by.

The breakthrough began in the 1930s. Geneticists found that mutations alone did not provide an adequate answer to the conundrum of evolution; natural selection, too, played a role in changing the frequencies of genes in populations. Population genetics and fruit fly experiments showed that most variation was due to a recombination of genes from both parents but that additional variation was the result of slight mutations. These random variants were then weeded out by natural selection. Selection by the environment affected the degree to which a particular gene could spread within a population; the stronger the selection, the more rapid the genetic change.

The outcome that emerged during the 1940s was a synthesis that combined Darwin's mechanism of natural selection with the developing understanding of the mechanism of heredity. It was known as the 'Neo-Darwinian synthesis' (or 'modern synthesis') and it put an end to the perennial feuds of rival schools of evolutionists. Evolution, it was agreed, was a gradual, long-term process consisting, essentially, of the accumulation within lineages of small genetic mutations and recombinations which, over long periods of time, resulted in large effects. This generation-to-generation modification of gene frequencies was guided by the hand of natural selection. These processes explained not only changes within species but also the origins of new species and biotic diversity.

Advances in genetic research added weight to the synthesis. In 1944, Oswald Avery and two colleagues established that genetic information is contained not in cell proteins, as had been thought previously, but in nucleic acids—in DNA (deoxyribose nucleic acid) molecules. Their discovery marked the start of the 'molecular revolution'. A further crucial breakthrough came in 1953, when James Watson and Francis Crick deciphered the double-helix structure of the DNA molecule in which genetic information is coded.

Other ideas incorporated into the synthesis, however, proved to be more controversial. Two of its principal architects, the geneticist Theodosius Dobzhansky and the ornithologist Ernst Mayr, considered the evolutionary history of humans to have been different from that of other organisms, proceeding upward in a straight line rather than by trial and error. Dobzhansky's argument was based on the advent of human culture, in particular toolmaking. Culture, he said, had removed humans from the effects of the natural environment and therefore from natural selection, enabling them to adapt to a diversity of environments. The process of acquiring culture, he argued, must have been confined to a single hominid species. He explained the differences that existed among living and extinct humans as being simply different adaptations of members of a single evolving lineage. 'As far as [is] known', said Dobzhansky, 'no more than one hominid species existed at any one time level'.

Mayr reached a similar conclusion, though from a different perspective. After several years of fieldwork on the birds of New Guinea and the Solomon Islands, Mayr became convinced that geographic features played a crucial role in the production of species. Speciation, he observed, occurred only through an initial phase of geographical isolation, breaking a once-homogenous population into subpopulations; these then began to evolve in different directions, according to the demands of their local environment. Mayr introduced the term 'allopatric speciation' to describe the process, meaning speciation 'in another place'. Small isolated populations on the fringe of a main population, separated by barriers such as mountains or stretches of water, were the most likely source of new species, he said. Once they had become distinct, they were no longer able to interbreed with the main population. In other words, geographical forces rather than biological factors were responsible for dividing a population. Speciation could only occur when there was a vacant ecological niche that a subspecies could occupy.

Mayr went on to argue that because humans were so geographically widespread and so diversely adapted, there had been few opportunities for speciation except, perhaps, at the very beginning. There was no justification, he said, for palaeontologists and palaeoanthropologists to designate large numbers of human types, as Robert Broom had done with the australopithecines. Mayr was ready to include *Australopithecus* as part of the hominid family. But he insisted initially that only one species of *Australopithecus* could have existed, on the grounds that according to the principle of competitive exclusion, if there were two rival species occupying the same territory, one would have driven the other to extinction. He eventually accepted that because of the degree of morphological difference, there must have been two species a robust form which had died out and the more lightly built *Australopithecus africanus*. It was *Australopithecus africanus*, he decided, that had evolved into *Homo*.

As a result of Mayr's strictures, scientists agreed on a new dispensation: australopithecines were accepted as the first member of the hominid family; next came *Homo erectus* a category said to have evolved from australopithecines that included *Pithecanthropus* from Java and similar fossils from China then came *Homo sapiens*. The Neanderthals were no longer regarded as an extinct primitive species but as an early variant of *Homo sapiens* perhaps adapted to the colder climate of the Ice Ages. How to fit Piltdown Man into this scheme remained unsolved. Thus, human evolution was envisaged as a straight line of continuous transformation of one species into the next.

Although the modern synthesis provided a more straightforward structure for understanding human evolution, it lost the diversity that palaeoanthropologists had hitherto enjoyed to accommodate the range of fossils they had discovered. A diverse assortment of fossils was now lumped together as variations on a theme because they originated from roughly the same period. Species became no more than segments of steadily evolving lineages. According to the modern synthesis, with time and anatomical change marching in tandem, the fossil record should have shown gradual intergradations from one species to the next. The difficulty was that fossil hunters had so far found too few hominid specimens to ascertain whether the theory fitted the facts on the ground.

In Africa, meanwhile, the Leakeys had found a benefactor who was to transform their fortunes. Responding to a letter that Louis Leakey wrote to the London *Times* in February 1948, explaining the difficulties and expense of organising field expeditions, Charles Boise, an American-born London businessman, sent him a cheque for £1,000. An art connoisseur, Boise had once worked in the Belgian Congo and was fascinated by prehistory and palaeontology. He jumped at the opportunity to assist Leakey's plans for an expedition to Miocene sites on Lake Victoria's Rusinga Island dating back 20 million years.

Leakey's interest in the potential of Miocene sites in western Kenya had been prompted by discoveries made there during the 1920s and 1930s. In 1927, a white farmer in Koru quarrying for agricultural lime had found some fossilised bones, including the upper jaw of an ancient ape, which were eventually sent to Arthur Hopwood at the British Museum for examination. Hopwood was convinced that the jaw belonged to an ancestor of chimpanzees. In 1931, he spent five weeks exploring Miocene sites in western Kenya, recovering nine more specimens of apelike creatures from Koru and numerous other mammal fossils. Writing in a London journal in 1933, Hopwood labelled the Miocene apes *Proconsul africanus*—'Before Consul'—taking the name of a famous zoo chimpanzee. The following year, Leakey had begun exploring Miocene deposits on Rusinga Island, finding sixteen more specimens of Miocene apes. On subsequent visits, he found several more fragments, including the most complete jaw of any Miocene ape yet discovered.

The Leakeys' 1948 expedition to Rusinga was a triumph. Two days after setting up camp under a spreading fig tree near Kaswanga Point, Mary spotted the glint of a tooth on the sloping surface of a small gully. Carefully brushing the surrounding area, she uncovered the greater part of a *Proconsul* skull, the first ever to be discovered. 'This was a wildly exciting find', she wrote in her autobiography. 'Ours were the first eyes to see a *Proconsul* face'.

The discovery made news around the world. When Mary took the skull to England for examination, she was met by a scrum of reporters.

The reception at the airport, she cabled to Louis, was 'overwhelming'. Crowds flocked to see the skull on exhibition at the Natural History Museum in London. Delighted by the find, Boise promptly provided a cheque for £1,600 to assist further expeditions to Rusinga. The tally of finds at Miocene sites in western Kenya eventually rose to more than 15,000 fossils, including bones of 450 individual apes.

In 1951, the Leakeys took Charles Boise to Olduvai. During previous visits they had managed to cover much ground, exploring in all 180 miles of exposures and marking out potential sites. But they had never been able to undertake any major excavations. Louis's great ambition was to find evidence of the early man he believed had produced the stone tools uncovered there. Although Boise found the heat, wind and dust tiring, he nevertheless enjoyed himself and set up a covenant for the Leakeys to carry out excavations for the next seven years. It was to prove to be a turning point in their work there.

In England, the demise of Piltdown Man finally came in 1953. The chain of events that ultimately exposed the fraud started when Kenneth Oakley of the Natural History Museum subjected the fossils to a new dating technique, the fluorine absorption test. The tests established that they were relatively modern. Subsequent investigations swiftly revealed the hoax. Every single one of some forty finds at Piltdown had been forged and planted there: The skull fragments belonged to a modern human, 600 years old; the jawbone belonged to an orang-utan estimated to be 500 years old; the teeth had been filed down to produce a human pattern of wear; all the objects had been carefully stained to give them an aged appearance. What was astonishing was that so many eminent scientists had allowed such a crude forgery to pass without challenge for so long.

Piltdown Man was no ordinary hoax: It was a systematic campaign carried out over several years. The early skull fragments were created in advance and salted with the intention that more extensive finds would be planted at a later stage. Pieces were put together to fit in with the prevailing view about what an ancestral human should look like. Convinced of its authenticity, some scientists—including Sir Arthur Smith Woodward—continued to work at the Piltdown site for years in the hope of finding more evidence. In 1948, on his deathbed, Woodward had dictated the text of a book entitled *The Earliest Englishman*.

A minor literary industry sprang up to identify the culprits. The list of suspects was long but was eventually whittled down to two men. The most obvious candidate was Charles Dawson, the local lawyer who was the first person to search for fossils in the Piltdown quarry. An amateur archaeologist, he was known to be keen to build up a reputation for himself in academic circles by making spectacular discoveries, but the full extent of his ambition did not become evident for decades. In subsequent investigations, Dawson turned out to be an inveterate hoaxer and forger, claiming dozens of significant finds that later turned out to be fraudulent.

The other candidate was Martin Hinton, an eccentric and devious character with a reputation for playing practical jokes who rose to become the Keeper of Zoology at the Natural History Museum. Throughout the period of the Piltdown affair, he was working as a curator in the museum's Geology Department. One possible motive attributed to him was that he was driven by an abiding dislike of the department's Keeper of Geology, Arthur Smith Woodward, and set out to discredit him. In 1978, contractors clearing a loft in the southwest tower of the museum found a trunk belonging to Hinton which turned out to contain bones stained and carved in the same way as the Piltdown fossils and associated artefacts. Researchers considered it to be 'a smoking gun'.

Whoever the perpetrator was, for British scientists it was a humiliating episode. Raymond Dart called it 'the most outrageous palaeontological fraud which was to mislead the thinking of every recognised authority in anthropology for forty years and so arrest the recognition of the Australopithecines until July 1953'.

Over the course of seven years, from 1952 to 1958, the Leakeys toiled away at Olduvai, spending short seasons there with a small team of assistants, working intensively until the money ran out. Concentrating on Bed II, they uncovered huge numbers of artefacts and animal fossil bones, and laid bare the ancient living floors of early hominids but found virtually no evidence of hominid remains; their haul amounted to just two teeth.

In June 1959, they decided to change the focus of their work to Bed I. Within a day, their veteran field assistant, Heselon Mukiri, spotted a hominid tooth. The area around it looked promising, but their funds for the season were almost exhausted. Louis hurried back to Nairobi to obtain an overdraft to cover three more weeks of excavation. He also arranged for a cameraman, Des Bartlett, who worked for a British television series, *On Safari*, to film the excavation from the start.

While waiting for Bartlett to arrive at Olduvai, the Leakeys spent a few days exploring other sites. On the morning of 17 July, they had planned to investigate an area in the gorge several miles from the camp. But Louis awoke with a fever, so Mary set out on her own, taking two dogs with her, deciding to explore a different site nearer to camp.

She found plenty of material there—broken stone tools and fragmented fossils littered the surface. But one scrap of bone caught her eye. Unlike other items, it was not lying on the surface but projecting from beneath it. 'It seemed to be part of a skull', she wrote. 'It had a hominid look, but the bones seemed enormously thick—too thick, surely'. Her doubts soon vanished.

> I carefully brushed away a little of the deposit, and then I could see parts of two large teeth in place in the upper jaw. They *were* hominid. It was a hominid skull, apparently *in situ*, and there was a lot of it.

Mary rushed back to camp. 'I've got him! I've got him!' she shouted. Louis was quickly on his feet, heading for the site with Mary. But the sense of exhilaration he felt when he saw the fossil was quickly overshadowed by disappointment. He had hoped to find a *Homo*—a fossil that would turn out to be the world's earliest known human. But the teeth looked similar to those of the australopithecines of South Africa. 'Oh dear', he said to Mary, 'I think it's an australopithecine'. Although Leakey knew that the find was highly significant, he was one of the few remaining scientists who held adamantly to the view that australopithecines were not 'true' ancestors of humans but an 'aberrant offshoot' from 'the stock which gave rise to man'. Writing in his diary of the day's events, he pondered over the outcome: 'I wonder what tomorrow will reveal. Is it a very primitive man or skull of an *Australopithecus*?'

In the following days, as fragments of the skull were slowly removed, Leakey became all the more convinced that it belonged to an early type of human. It had a mixture of features: a flat face, a flat nose, wide cheekbones, a massive jaw, big back teeth, tiny front teeth, hardly any forehead and a bony crest running the length of its skull. But what ultimately persuaded Leakey was that the skull had been found in an area replete with stone tools. The common view at the time was that toolmaking defined the boundary between human and pre-human; it was said to be the hallmark of humankind. Like most scientists, Leakey believed that australopithecines, because of their small brains, had been incapable of making tools. His assumption was that Olduvai's stone tools could only have been made by an early type of human and that therefore the skull that had been found had to be human.

On his return to Nairobi in August, Leakey lost no time in sending an article to *Nature* announcing that he had found 'the oldest maker of stone tools' yet discovered. While acknowledging that the skull bore some resemblance to australopithecines, he argued that the differences warranted placing it in a separate genus. 'I therefore propose to name the new skull *Zinjanthropus boisei*'.

The name was a concoction of *Zinj*, an old Arabic word for the coast of eastern Africa; *anthropos* from the Greek word for man; and *boisei* in honour of Charles Boise, the Leakeys' loyal benefactor. But the Leakeys themselves referred to the skull more affectionately as 'Dear Boy'.

'Zinj' became one of the most famous fossils ever discovered. Soon after returning to Nairobi, Leakey embarked on a world tour to show it to admiring audiences. His address to the fourth Pan-African Congress on Prehistory in Léopoldville, capital of the Belgian Congo (now Kinshasa) at the end of August, with Zinj balanced next to him on the podium, created a sensation. Among the delegates were Raymond Dart and his young protégé Phillip Tobias, the newly appointed Professor of Anatomy at the University of the Witwatersrand. 'There was tremendous applause', recalled Tobias, 'and not just a murmur of conversation, but quite hysterically excited conversation from everyone present'. Impressed by the size of its jaw and teeth, Tobias remarked: 'I have never seen a more remarkable set of nutcrackers', providing the press with an instant nickname: Nutcracker Man.

The occasion was not quite the triumph that Leakey wanted, however. Many scientists observed that Zinj possessed strong similarities to South Africa's 'robust' australopithecines, as Leakey himself had initially thought, and they drew the opposite conclusion to Leakey about the presence of stone tools: Instead of proving that Zinj was human, it meant that small-brained australopithecines might have been capable of making tools. Furthermore, they were critical of Leakey for adding yet another genus and species to the list, with seemingly little justification.

Nevertheless, excitement about Nutcracker Man spread across the globe. In London, Leakey and Zinj were treated as celebrities. A talk that Leakey gave to the British Academy in Piccadilly was packed. 'Television stars like Sir Mortimer Wheeler [a prominent archaeologist] had to stand against the wall', reported one newspaper. Leakey was greeted with similar acclaim during a marathon tour of the United States. Over the course of a month, he delivered sixty-six lectures at seventeen universities and other scientific institutions, inspiring audiences with his story of searching for early man for thirty years, persevering against all odds. After a presentation he made in Washington, DC, the National Geographic Society decided to award the Leakeys a grant of $20,000 to continue their work at Olduvai—the largest sum of money they had ever seen.

Having made its debut on the world stage, Zinj was taken to Johannesburg for a thorough evaluation, a task given to Phillip Tobias. He eventually concluded that Zinj was an australopithecine but one that merited a new species that he called *Australopithecus boisei*. The picture that emerged was of a stocky creature, robustly built, with wide cheekbones, mostly bipedal, but with a relatively small brain—no more than 530 cubic centimetres.

The full significance of Zinj, however, only came to light in 1961. In calculating its age, Leakey had estimated Zinj to have lived 'more than 600,000 years ago', a figure based on his assumption about the age of the volcanic ash in Bed I where the skull had been found. But such estimates were regarded at the time as amounting to little more than guesswork. No practical way of determining the absolute age of volcanic material and the ancient fossils they contained had yet been devised. Although radiocarbon-dating techniques had been in use since the early 1950s, they had proved to be unreliable when applied to material beyond about 30,000 years old.

In the late 1950s, however, a geophysicist from the University of California at Berkeley, Jack Evernden, developed a new method of radiometric dating known as potassium-argon dating. Evernden twice visited Olduvai to collect samples of tuffs and basalt; and his colleague Garniss Curtis took further samples in 1961. The results of their tests were announced in *Nature* in 1961. To general astonishment, they showed that the basalt in Bed I—and hence the age of Dear Boy—was around 1.75 million years old.

CHAPTER 6

Handy Man

A GOLDEN AGE of exploration followed the discovery of Zinj. The scientific community was by now generally agreed that Africa rather than Asia was the most likely birthplace of humankind. Inspired by Louis Leakey's publicity campaign and by television programmes about Zinj, popular interest in human origins soared. A new generation of students took up the cause of palaeoanthropology. Funds for more fieldwork were readily available.

At Olduvai, the Leakeys were able to construct a permanent camp for the first time, to recruit full-time African staff and to invite specialist scientists to assist them. Mary spent most of her time there directing excavations, making only short visits to Nairobi. Louis was preoccupied with his work as curator of the National Museum in Nairobi and with numerous other projects and joined her for weekends and vacations. Each year he conducted long lecture tours in the United States, thrilling American audiences with impassioned accounts of his work on human origins and raising more funds. 'As the years passed', wrote Mary in her memoirs, 'his reception in at least some parts of the States turned to outright hero-worship'.

The 1960 season at Olduvai proved highly productive. In May, the Leakeys' nineteen-year-old son, Jonathan, wandering off on a fossil hunt of his own, discovered a fragment of the mandible of a sabre-toothed cat lying on the surface about 100 yards from the Zinj site. Hoping to find further remains, Jonathan sieved through surface deposits. Nothing more of the cat turned up, but he found instead a hominid tooth and toe bone. With growing excitement, a new excavation was started at what became known as 'Jonny's site'. In August, Mary uncovered fourteen foot bones there—the first discovery of the foot of an early human. In the following weeks, 'Jonny's site' yielded remains from several different specimens—hand bones, parts of a skull and a lower jaw complete with thirteen opalescent teeth. The skull bones appeared distinctly different from those of its neighbour Zinj: They were thinner; the braincase was larger; there was no sagittal crest.

Assessing the evidence, Louis Leakey became increasingly convinced that the fossils from 'Jonny's site' represented a new species, a primitive *Homo*, possibly the toolmaker he had long sought. He was also struck by the similar age of the two sites. The fossils at 'Jonny's site' had come from deposits one foot lower than the Zinj deposits, making them slightly older but contemporary. What this meant, Leakey believed, was that 'two entirely distinct hominids' had lived together at Olduvai, 'side by side'.

Determined to avoid controversy, Louis compiled a straightforward report of the new discoveries for publication in *Nature* in February 1961, refraining from speculation. He gave a similarly cautious account at a press conference at the headquarters of the National Geographic Society in Washington in February. The skull, he said, belonged to a juvenile about eleven years old. Because it had lived many thousands of years before *Zinjanthropus*, it was known simply as 'pre-*Zinjanthropus*'. He went no further than to say that it seemed to be 'a quite distinct type of hominid'.

But when a reporter asked him what had caused a hole in the skull and the fracture radiating from it, Leakey could not resist a bit of speculation and opened up a new chapter of controversy. The child had died, he suggested, as the result of an injury. 'I think we can take it for granted that the child was hit on the head by a blunt instrument. It was murder most foul'.

Newspapers around the world seized on the idea, running headlines proclaiming 'World's First Murder'. In England, the scientific establishment, noting that Leakey's account in *Nature* had made no mention of a blow to the head, reacted with disdain. A correspondent for the London *Times* reported: 'British anthropologists were left wondering during the weekend what new consideration of evidence had led Dr L.S.B. Leakey . . . to bring in a verdict of murder, hundreds of thousands of years after the event'. The *New Scientist* rebuked Leakey for indulging in 'wild speculations' and accused him of making an important field of science look 'more than a little ridiculous'. The magazine *Punch* ran a satirical article entitled 'More Secrets from the Past: Oboyoboi Gorge', featuring the exploits of a well-known anthropologist Dr C.J.M. Crikey.

To help him establish a proper identity for 'pre-*Zinjanthropus*', Leakey sought the opinion of a number of other experts. Among them was Wilfred Le Gros Clark, Britain's leading palaeoanthropologist. From the outset, Le Gros Clark was doubtful about Leakey's claim about a new hominid; from all the evidence he had seen, he told Leakey in June 1961, he considered the fossils to be inseparable from *Australopithecus*. The South African anatomist Phillip Tobias took a similar view. 'My present feeling about the child is that it is an australopithecine', he wrote in May 1962 after studying the skull.

Leakey insisted, however, that 'pre-*Zinjanthropus*' was not an australopithecine. 'Mary and I are sure (more and more so every time we go over the data) that it is NOT *Australopithecus*', he wrote to Tobias in December 1962. Instead, he argued, it was 'a very primitive *Homo*'.

But Tobias continued to hold out. A major obstacle he faced was the size of the brain. The scientific consensus at the time was that a hominid needed to exceed a certain brain size to qualify for the genus *Homo*. According to Le Gros Clark, a large brain was a 'distinctive human trait'. Although there was no agreement among anatomists about a specific threshold, the general size they settled on ranged from 700 to 800 cubic centimetres.

The first estimate that Tobias gave for the 'pre-*Zinj*' brain size was between 600 and 700 cubic centimetres. He told Leakey that this made it 'difficult to reconcile with *Homo*'. But Leakey would not relent. 'Phillip took a lot of persuading', Mary Leakey recalled. 'Louis had to bludgeon Phillip to convince him. No one lightly names a new hominid species. But Louis loved it'.

The turning point came in 1963 when Mary found further specimens that appeared to be related to 'pre-*Zinj*'. In all, Olduvai yielded the remains of eight 'pre-*Zinj*' hominids, providing sufficient evidence for Tobias to change his mind. 'They all had bigger brains and narrower teeth and a number of other features . . . which showed a nearer approach to the genus *Homo* than to *Australopithecus*', he explained later. Nevertheless, the final estimate for the brain size that Tobias produced—about 680 cubic centimetres—meant that 'pre-*Zinj*' still fell short of the accepted threshold for human membership.

Other experts whom Leakey contacted provided further support for his claims. John Napier, a hand specialist at the Royal Free Hospital in London, concluded that the 'pre-*Zinj*' hand bones displayed modern features, including a long opposable thumb. Its hands, he said, were capable of two types of grip, a power grip and a precision grip, both essential requisites for a toolmaker. Such a hand, he noted, would have had the 'physical capacity' to make the small pebble-tools found at Olduvai. The foot bones added to the evidence. Examined by Michael Day of the Royal Free Hospital, they indicated that 'pre-*Zinj*' was bipedal, walking upright, not just occasionally but habitually.

In sum, the corpus of evidence that Leakey assembled depicted a small, slenderly built creature, with a relatively large brain, larger than any australopithecine, humanlike teeth and hands capable of making tools. It lived nearly 2 million years ago; and, according to Leakey, was a direct precursor of modern man.

Announcing their findings in a joint paper published in *Nature* on 4 April 1964, Leakey, Tobias and Napier set off a storm of controversy. For in order to incorporate 'pre-*Zinj*' within the genus *Homo*, they needed to rework the definition of *Homo*.

'We have come to the conclusion that, apart from *Australopithecus* (*Zinjanthropus*), the specimens we are dealing with from Bed I and the lower part of Bed II at Olduvai represent a single species of the genus *Homo* and not an australopithecine', they wrote. 'But if we are to include the new material in the genus *Homo* (rather than set up a distinct genus for it, which we believe to be unwise), it became necessary to revise the diagnosis of this genus'.

They set the new cerebral Rubicon at 600 cubic centimetres. 'The cranial capacity is very variable but is, on average, larger than the range of capacities of members of the genus *Australopithecus*, although the lower part of the range of capacities in the genus *Homo* overlaps with the upper part of the range of *Australopithecus*; the capacity is (on average) larger relative to body-size and ranges from about 600 c.c. in earlier forms to more than 1,600 c.c.'

This new species of human was given the name *Homo habilis*. The name had been devised by Raymond Dart to describe what was said to be the world's first toolmaker, someone who was 'able, handy, mentally-skilful, vigorous'—a 'handy man'.

Although Leakey had previously held firm to his belief that *Zinjanthropus* had been the 'earliest known stone-tool making man', he now agreed that *Zinjanthropus* was no more than an australopithecine probably incapable of making tools. But by switching his allegiance with such nonchalance from *Zinjanthropus* to *Homo habilis* as the toolmaker, he endured severe criticism.

There was even harsher condemnation of his arbitrary attempt to make radical alterations to the accepted definition of *Homo* in order to shoehorn 'pre-*Zinj*' into it. Le Gros Clark continued to insist that 'pre-*Zinj*' was an australopithecine. 'One is led to hope that [*Homo habilis*] will disappear as rapidly as he came', wrote Le Gros Clark. Critics also accused Leakey of confusing cultural notions about 'man the toolmaker' with morphological evidence. They pointed out that even though stone tools had been found with both *Zinjanthropus* and *Homo habilis*, Leakey had seen fit to claim that it was 'probable' that *Homo habilis* had been the 'more advanced toolmaker'—without producing any evidence. He had merely assumed that because *Homo habilis* appeared to have a larger brain, it was the more likely candidate to have made the tools.

Leakey was unrepentant. Addressing a gathering in Washington, DC, he declared:

> To me the most significant step that ever was taken in human history, the thing that turns animal into man was this step of making tools to a set and regular pattern. This is why we chose that definition of *Homo* . . . Once he had made the simplest of tools, he immediately opened himself a completely new food supply—and enhanced his chances of competing with other creatures.

And in a later press release, he urged colleagues 'to review all their previous ideas about human origins and to substitute for those theories new ones which were more in keeping with the facts that are now known'.

Although reaching his sixties and suffering from arthritic pains in his hip joints and other ailments, Louis Leakey kept up a frenetic pace of activity. As well as establishing a Centre for Prehistory and Palaeontology in Nairobi, he took an increasing interest in primate research. He fostered the careers of three young primatologists—or 'ape ladies', as they were sometimes called—who went on to achieve worldwide fame: Jane Goodall, whom he despatched to study chimpanzees in the forests bordering Lake Tanganyika; Dian Fossey, whom he sent to the Virunga Mountains to study highland gorillas; and Birute Galdikas, whom he helped to study orang-utans in Indonesia. He was also instrumental in opening a new chapter in African exploration.

At a luncheon given by Kenya's leader, Jomo Kenyatta, at State House in Nairobi in 1965, Leakey met Ethiopia's Emperor Haile Selassie, who asked him why fossils had been found in Tanganyika (Tanzania) and Kenya but not in Ethiopia. 'Well, Your Royal Highness', replied Leakey, 'if you would allow us to go and search in your country, I know where we might find something'. Haile Selassie wanted to know why he had not already been there to look. 'Well', replied Leakey, 'it's always been difficult. Your government has not given us the facilities'. 'All right', said Haile Selassie, 'I'll arrange it'.

CHAPTER 7

KOOBI FORA

RISING IN THE Ethiopian highlands, the Omo River runs southwards for 700 miles towards the Kenya border before flowing into a landlocked desert lake known as the Jade Sea. = Turkana A French aristocrat, Robert Bourg de Bozas, discovered fossiliferous deposits in the lower Omo Valley in 1902, but it was not until 1932 that an expedition led by the French palaeontologist Camille Arambourg went to investigate the area. Arambourg's team took several tons of vertebrate fossils back to Paris but found no evidence of hominids. In 1959, a thirty-four-year-old American palaeoanthropologist, Clark Howell, spent several weeks searching the Omo deposits, but the fossils he found were confiscated by Ethiopian border guards on his departure and the Ethiopian authorities subsequently refused him permission to return to the Omo.

It was to this remote, arid stretch of southern Ethiopia—the homeland of nomadic, warring tribes—that Haile Selassie now agreed that Leakey should mount an international expedition. A local wildlife warden warned: 'This bit of Africa has never been governed or administered by anyone at all and primitive reactions prevail'. Leakey was given nominal charge of the venture, but the pain he suffered from arthritic hips prevented him from participating in any fieldwork and he planned only to make occasional visits. Instead, Leakey arranged for Arambourg, then aged eighty-two, to lead a French team; and for Clark Howell to lead an American team. At Howell's insistence, the teams included geologists, anatomists, archaeologists and other specialists skilled in examining the fossils of plants and animals as well as hominids. Howell's aim was to turn the field itself into a laboratory that would help reveal the wider world in which hominids lived. His multidisciplinary approach—in marked contrast to the popular image of the lone field scientist that Leakey cultivated—set a new standard for palaeoanthropological research.

Added to this array of talent was a team from Kenya with less expertise. Because of his failing health, Leakey asked his twenty-three-year-old son Richard to lead it. Richard Leakey had hitherto shown little interest in following 'the Leakey tradition'. He had dropped out of secondary school at the age of sixteen without any qualifications and had become involved in the safari business in East Africa, organising supplies, transport and personnel for tourists and film crews. 'I was determined to distance myself from my parents and their work on fossils and prehistory, largely because I wanted to be my own man', he recalled. Louis had hired him essentially as a safari manager, relying on him to deal with the logistics of the expedition. The only member of the Kenyan field team with professional training in archaeology and anthropology was Richard's wife, Margaret. On scientific matters Richard was expected to keep in radio contact with his father.

The Omo Research Expedition set out from Nairobi in June 1967 in a convoy of trucks and Land Rovers carrying forty people, taking five days to cover the 500-mile journey to the lower Omo Valley. It was billed as an international enterprise but rapidly divided into three competing groups. Each team set up camp in a separate concession area, determined to become the first to find the oldest, most spectacular fossil. The French had been allocated the southernmost fossil deposits where Arambourg had previously worked; the Americans were

1

The British naturalist Charles Darwin speculated that the origins of humankind were likely to be found in Africa, but his theory was generally dismissed for half a century.

2

In one of his journals, Darwin sketched a "Tree of Life" with multiple branches, a prescient idea that eventually proved to be accurate.

3

Ernst Haeckel, an influential German biologist, argued that Asia, not Africa, was the most likely birthplace of humankind.

4 HAECKEL'S EVOLUTION OF MAN. *PLATE XV.*

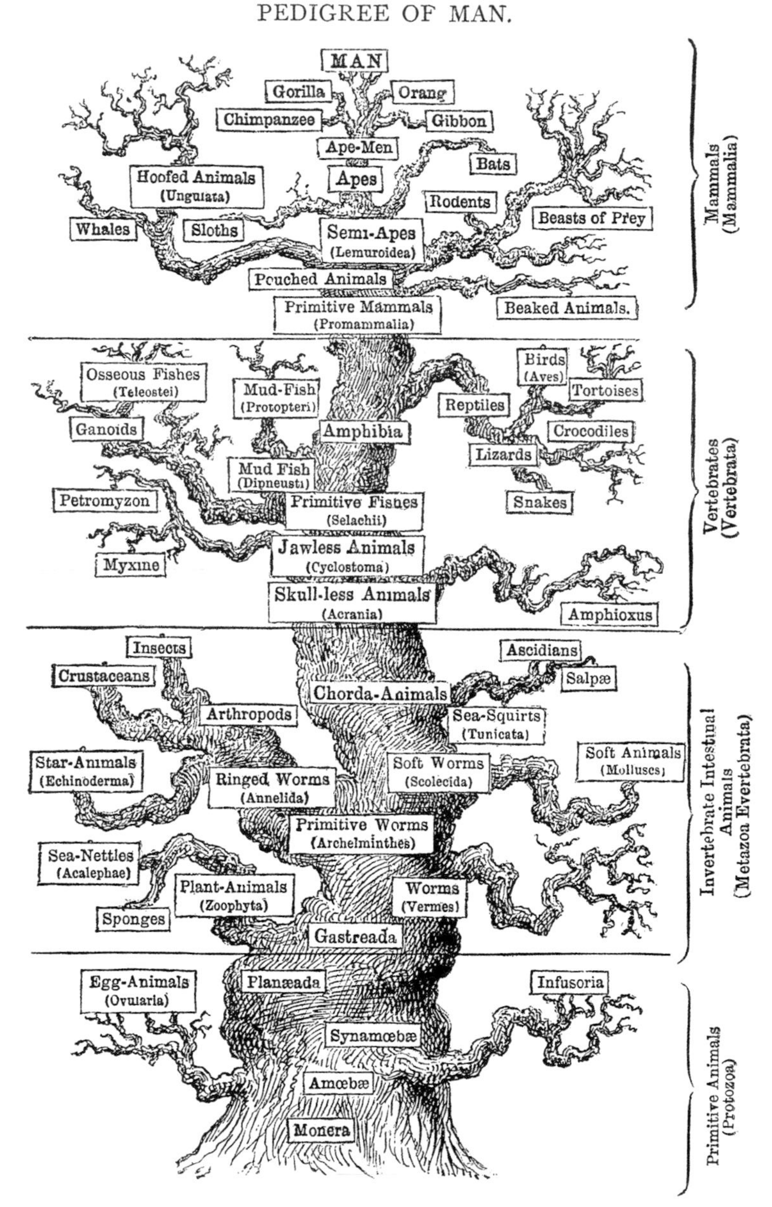

Haeckel's Tree of Life differed from Darwin's. It showed a main trunk leading upward from 'lower' forms towards humankind at the pinnacle of creation.

Raymond Dart, an Australian scientist whose claims that a hominid skull known as the Taung child came from a line of proto-humans was ridiculed by the British scientific establishment.

5

Robert Broom, a Scottish-born palaeontologist, at work in the field in South Africa. His discoveries supported Dart's claims and helped establish australopithecines as a significant category of human ancestors.

6

7

Mary and Louis Leakey at work in the field at Olduvai Gorge, Tanzania. It was not until 1959, after nearly 30 years of searching at Olduvai, that Mary made their first major hominid discovery there: a skull belonging to a 1.75 million-year-old australopithecine that the Leakeys named *Zinjanthropus boisei*, but it was more generally known as Nutcracker Man because of its huge teeth.

8

Mary Leakey pictured beneath a cliff at Olduvai in 1955.

Expeditions to Olduvai were always a family affair that included Mary's beloved Dalmatians. Their son Philip also joined in the work, making discoveries of his own.

11

Louis Leakey pictured with *Zinjanthropus* and friends.

12

Mary Leakey pictured in 1972.

13

Fossil-hunters at work at Olduvai.

Ol doinyo Lengai. The Maasai's 'Mountain of God', the only volcano still active on the eastern branch of Africa's Great Rift Valley.

15

Olduvai Gorge.

16

Camels at Koobi Fora. Richard and Meave Leakey, Kamoya Kimeu and Peter Nzube ride out into the desert in search of fossils. The experiment with camels was largely impractical and soon abandoned.

17

Inspecting fossils at Koobi Fora, with Kamoya Kimeu (*top right*).

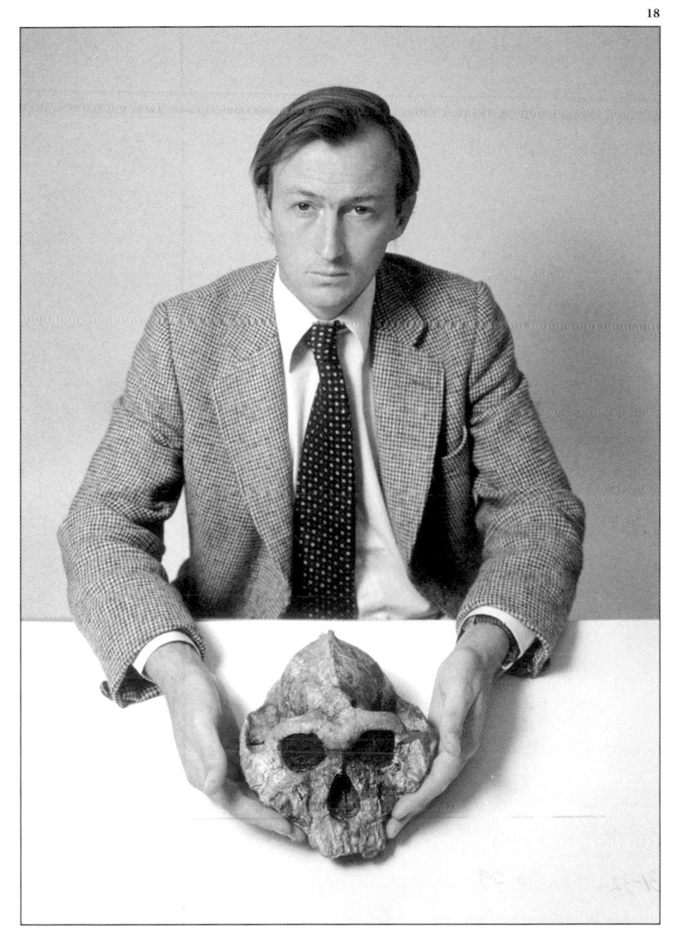

Richard Leakey with a complete skull of an *Australopithecus robustus* he found at East Turkana in 1969.

19

Richard Leakey holding two prized hominid specimens: a complete skull of an *Australopithecus robustus* and a skull known as 1470 that became the subject of a fierce controversy.

pp. 35 and 31-32 p. 79

20

Alan Walker, a British-born palaeontologist involved in several key discoveries in the Turkana region of Kenya.

21

Donald Johanson points out features of 'Lucy', an australopithecine dating back 3.2 million years ago that he found in 1974 in the Afar region of Ethiopia.

22

Donald Johanson and Richard Leakey held a joint press conference in March 1976 to explain their recent hominid discoveries. They were later to become bitter rivals.

23

Tim White (*far left*) meets Donald Johanson for the first time in January 1976 in a museum laboratory in Nairobi. Also present are Richard Leakey and Bernard Wood, a British anatomist. Johanson had just arrived from Hadar, Ethiopia, with his collection of fossils known as the First Family.

sited further to the north; the Kenyans got the least interesting area of deposits—fifty square miles of sediments that lay on the opposite side of the Omo River. Richard Leakey soon decided that both the French and the Americans were lazy and incompetent; and the French and Americans considered him to be untrained and uneducated.

It was the Kenyans who nevertheless turned up the first significant find. Three weeks after their arrival, a sharp-eyed Kamba fossil-hunter, Kamoya Kimeu, who had worked with the Leakeys at Olduvai since 1960, spotted fragments of a hominid skull in rock formations on the banks of the Kibish River; excavations produced other fossil parts. Upon examination, Omo I, as it was called, proved to be a specimen of modern man—*Homo sapiens*—but one that was far older than any previous discovery. The prevailing view at the time was that *Homo sapiens* had emerged about 60,000 years ago. Omo I was estimated to be 130,000 years old, though, because of the difficulties of dating with accuracy, there were doubts about whether this estimate was reliable.

The Kenyan discovery, however, was soon overshadowed by a French triumph. Two weeks later, Yves Coppens from the Musée de l'Homme found the lower jawbone of a primitive australopithecine dating back at least 2 million years. By the end of the season, both the Americans and the French had made further gains, but the Kenyans had little to show apart from another 'modern-man' skull. 'I was conscious that my team was doing less well than the others', Richard Leakey wrote in his autobiography *One Life*.

Moreover, Leakey had become increasingly frustrated at being treated, as he put it, like 'a sort of tea-boy'. His father's occasional appearances served only to emphasise his junior status. Recalling the expedition years later, Richard Leakey told Delta Willis: 'I really didn't like the Omo project. There were too many chaps ahead of me—all sorts of serious, senior scientists. I was very much on the bottom of the pile'. Ambitious, arrogant and ruthless, what he wanted was control of his own expedition.

A chance event provided him with an opening. On his way back to the Omo Valley by plane after a short trip to Nairobi in August, a huge storm system lay across the normal route along the western shore of the Jade Sea—Lake Turkana—so the pilot diverted to the east side, an area that Leakey had never seen before. Maps showed the region to consist of nothing more than volcanic rock, but from the window Leakey saw what looked like vast stretches of sandstone sediments traversed by eroded gullies, terrain likely to produce fossils. A few days later, borrowing a helicopter that the Americans had chartered, he set off on a reconnaissance trip. Spotting some promising sedimentary outcrops, Leakey asked the pilot to land. 'I jumped out of the helicopter and immediately found fossils as well as stone artefacts', Leakey wrote in his autobiography. 'We explored further, landing at various other outcrops and everywhere we stopped there were fossils'. He was so excited by the finds that he forgot to note their location. But his mind was now made up: He would return to Lake Turkana the following year to lead his own expedition.

To do so, Richard had to outflank his own father. Keeping his intentions secret, he accompanied Louis to a meeting in Washington of the National Geographic Society's Research and Exploration Committee, the Leakeys' main benefactor. The meeting was scheduled to hear reports on the Omo expedition and on other projects in East Africa and to consider their future funding. Louis duly described his plans for the coming season at Omo and Olduvai, and in the case of Omo, asked for $25,000 to enable the Kenyan team to continue its work there in 1968. Richard followed with his own account of the Omo expedition. But at the end of it—to Louis's surprise—he brazenly suggested that instead of giving $25,000 to the Omo venture, National Geographic would do better to use the money to investigate sites on the shores of Lake Turkana that he had identified. For a moment, there was silence in the room. Then Louis spoke up, sighing over the impatience of youth and pointing out that there was already known to be an abundance of fossils at Omo awaiting searchers. The two Leakeys were asked to leave the room while the committee members deliberated over the matter. Behind closed doors, they decided to reward Richard's 'cheek and initiative'. Louis was given more funds for research at Olduvai and at a site in Kenya he had developed. But Richard gained $25,000 for his own expedition to Turkana. 'You can have the money', Melville Grosvenor, the society president, told him, 'but if you find nothing you are never to come begging at our door again'.

The outcome was a serious blow to Louis. He had initiated the Omo project but was now cut off from it. The French and American teams were left to continue the work there on their own, without Kenyan participation. He faced other woes. His health continued to deteriorate rapidly; his arthritis had worsened and he needed crutches or walking sticks to support him. He became irritable, constantly quarrelling with colleagues who disagreed with him. Many thought his scientific judgement was impaired.

His marriage to Mary, too, was failing. Like other scientists, she had begun to lose her professional respect for him. She had always been a more dedicated scientist than Louis, prepared to search patiently for evidence rather than to rush headlong into speculation, seeking the limelight. After closing down her Olduvai excavations in 1963, she had spent five years meticulously analysing the huge collection of material retrieved from the lower beds. Among the items were 37,127 artefacts; nearly 20,000 animal remains; and twenty fossilised hominids. The artefacts included the oldest known tools, dating back 2 million years. Her magisterial volume on the Olduvai excavations was supposed to have been written in collaboration with

Louis, but Louis was invariably distracted, gallivanting after other projects and ideas. 'I had to watch Louis decline from the height of his intellectual powers and the fullness of his charm, to become irritable and irrational', she wrote in her autobiography. Mary was also weary of his many friendships with attractive young women. In 1968, she decided to resume work at Olduvai, living there permanently. They subsequently met for only brief periods.

The next generation was ready to make its mark. 'Richard was always a competitive person', Mary observed. 'When he entered a new field it was with the intention of getting to the top, and the sooner the better. And who at the time was in possession of the summit, needing therefore in due course to be replaced? Louis'.

The area that Richard Leakey marked out as his research territory in 1968 was a 500-square-mile stretch of volcanic badlands, extending northwards from Allia Bay to the Ethiopian border and inland from the lakeshore for some twenty miles. Much of it resembled a lunar landscape, a boundless expanse of lava and sand littered with the wrecks of ancient volcanoes. The winds and the heat were ferocious.

Only a small team set out on the first expedition. It consisted of a handful of young scientists willing to throw in their lot with Leakey and a group of Kamba fossil-hunters trained at Olduvai. No one knew what to expect. Leakey had no idea of the real conditions at Lake Turkana; nor did he have a clear plan to follow.

From the outset, however, the team found an abundance of animal fossils. During three months' exploration, they also turned up three hominid specimens—australopithecine jawbones. 'I realized then', wrote Leakey, 'that we had stumbled on to something far bigger and far more important than anyone suspected'.

The National Geographic Society was suitably impressed with the results and agreed to provide funds for further exploration. The team that Leakey assembled for his 1969 expedition was once again a small one consisting of fossil-hunters and scientists. Led by Kamoya Kimeu, the fossil-hunters soon became known as the Hominid Gang. Among the scientists there was a notable newcomer. Meave Epps was an English zoology graduate who had first arrived in Kenya in 1965 after responding to an advertisement that Louis Leakey had placed in the London *Times* seeking an assistant to work at a primate research centre he had established at Tigoni near Nairobi. She had spent two years at Tigoni, helping to run the centre and undertaking research on modern monkey skeletons, before returning to university to complete a doctorate. Although she did not meet Richard Leakey during her first visit to Kenya, she had heard much gossip about how brash, arrogant and unpleasant he was. 'I was always told, "You don't want to meet him. Meet any of the other Leakeys, but not Richard. You don't want to meet him"'. She returned to Kenya in 1968, invited by Louis Leakey to help out as acting director of the Tigoni research centre, and shortly after her arrival met Richard for the first time, finding him far more agreeable than his reputation suggested. The attraction was mutual. Richard's marriage to Margaret had fared badly and he was seeking a divorce. When Richard asked Meave Epps to join him on his second expedition, they were already lovers. They married the following year.

The site that Leakey chose for a base camp in 1969 was a sandy spit known as Koobi Fora that juts out for nearly a mile into Lake Turkana, ten miles north of Allia Bay. Surrounded by the lake on three sides, it was blessed by cool breezes blowing off the water. At first, Leakey's camp consisted of a few tents and a grass hut that was used as a laboratory, but in time he established a permanent base there, using local flagstones as building material. The lake itself was ideal for swimming but had to be shared with crocodiles and hippos.

With romantic notions of himself as a heroic explorer riding across the African desert, Leakey hired a collection of camels. The idea was largely impractical. The camels proved difficult to control, and by the end of each day, Leakey and other riders were left 'absolute wrecks'. The camels nevertheless provided him with spectacular photo opportunities. Like his father, Richard Leakey was adept at publicity and self-promotion, ensuring that press photographers were on hand to make the most of his exploits. The camel trips were soon abandoned. But the image of Leakey astride a camel heading out across the desert endured.

Aside from such antics, Leakey's second expedition to the Turkana basin was as successful as his first. Walking down a dry stream bed, he suddenly saw lying in the sand a complete fossil skull that bore a distinct resemblance to *Zinjanthropus*—the robust australopithecine skull that Mary Leakey had found at Olduvai ten years before; it had been washed out from the surrounding sediment during previous rains. A few days later, a member of the Hominid Gang found parts of another skull in the same area.

The team also came across evidence of ancient stone tools. While collecting samples of volcanic ash for dating purposes, a young American geologist, Kay Behrensmeyer, discovered a number of lava flakes lying on the surface of the 'tuff'—a layer of consolidated ash. An excavation at the site, which became known as the Kay Behrensmeyer Site, or KBS, revealed more stone tools. The tools came from a layer that Leakey estimated was at least as old as the oldest layer at Olduvai—about 1.9 million years old. Samples of the tuff sent to England for analysis, however, suggested a date of 2.6 million years. This meant that the tools at KBS were the oldest that had ever been discovered. Leakey was naturally delighted by the finding. But it marked the start of a controversy that plagued the scientific community for a decade.

By the end of the second expedition, the volcanic badlands of East Turkana were considered to be one of the most promising fossil sites ever discovered. Impressed by Clark Howell's multidisciplinary approach at Omo—where American and French scientists were still at work—Leakey set out to establish a similar research project based at Koobi Fora, recruiting a wide range of specialist scientists. His principal role became acting as their coordinator. With each new season, the tally of fossil finds soared—skulls, jaws and limb bones. In 1970, it included sixteen hominid specimens; in 1971, twenty-six.

The 1972 season produced an even more spectacular result. Stopping to examine a small pile of fragmented fossils on the surface of a steep gully, a novice member of the Hominid Gang, Bernard Ngeneo, (who had been taken on at Koobi Fora as a cook's assistant) thought he saw a piece of hominid bone. Other members of the search team had previously noticed the same pile but had assumed the fragments were antelope remains and had not bothered to investigate. After weeks of sieving, some 150 broken pieces of a hominid skull were recovered from the site. The task of piecing them together was undertaken by Leakey's wife, Meave, and a British-born anatomist, Alan Walker. Both possessed a particular aptitude for such painstaking work. As children, growing bored with ordinary jigsaw puzzles, they had both taken to making them more difficult by turning the pieces upside down, hiding the picture. Duly reassembled, the skull appeared to have a cranium capacity of about 800 cubic centimetres, far larger than the brain size of any known australopithecine and also Louis Leakey's *Homo habilis*. Moreover, it seemed to be far older. From its stratigraphic position below the KBS tuff, its age was estimated as at least 2.6 million years, possibly as old as 2.9 million years. It was given no name but was known by its catalogue number: 1470.

A difference of opinion soon emerged at Koobi Fora about the status of 1470. Alan Walker observed that apart from its large brain, it differed little from *Australopithecus africanus*. Leakey disagreed. He was convinced that it was a species of *Homo*—and therefore represented the world's 'earliest known man'. To avoid the kind of controversy that had erupted over *Homo habilis*, in an article in *Nature*, Leakey attributed it to an indeterminate species of *Homo* and called it *Homo* sp. indet.

In jubilant mood, Leakey took the skull to Nairobi to show his father, knowing how pleased he would be to see such evidence supporting his long-cherished belief in human antiquity. Their relationship had been strained for many years, but this was a cause for celebration. Louis's whole face lit up. A *Homo* skull with a large brain dating back more than 2 million years vindicated his decision to establish a new human species for 'pre-*Zinj*'—*Homo habilis*. Furthermore, it helped validate his contention that early humans and australopithecines had lived a contemporary existence, making it unlikely that australopithecines were a human ancestor. 'It's marvellous', he remarked, adding with a laugh, 'but they won't believe you'.

The following day, Louis Leakey left Nairobi to travel to the United States for yet another gruelling tour of lectures and fund-raising, but on reaching London he suffered a massive coronary and died there on 2 October 1972. Tributes poured in from around the world. One newspaper described him as 'The man with the million year mind'.

A month later, when Richard Leakey took 1470 to London, he became a media star. The fossil skull made newspaper headlines around the world. It was acclaimed the world's earliest known man, a million years older than the previous incumbent. A new Leakey era was born.

Within the scientific community, however, 1470 continued to stir controversy. Doubts were raised about the age of the skull. A South African palaeontologist, Basil Cooke, was the first to detect flaws in the dates claimed by Leakey's team for the KBS tuff. An expert on pig fossils, Cooke had discovered that because of the distinct changes that had occurred during the evolution of their molar teeth, fossil pigs could be used as a palaeontological 'clock' to check the work of geologists. From studies he had made at Olduvai and at the Omo River Valley, Cooke had estimated the age of a certain type of fossil pig found there—*Mesochoerus*—to be about 2 million years. Invited by Leakey to examine Koobi Fora's fossil pigs, Cooke found that the same type of pig taken from below the KBS tuff had been dated by Leakey's team as at least 2.6 million years. The Koobi Fora site and the Omo site were less than 100 miles apart. There seemed to be no logical reason for such a discrepancy in the evolutionary development of the same species. The possibility emerged, therefore, that Leakey's dates were at fault and that the 1470 skull and Koobi Fora tools which he had claimed to be the world's oldest were no older than the oldest hominids and tools from Olduvai—2 million years.

The initial dating of the KBS tuff had been carried out in England by two British experts, Jack Miller, a geophysicist at Cambridge University and Frank Fitch, a London-based geologist, using a new variant of the potassium-argon method that had astonished the scientific world with its revelation of Dear Boy's age. When queried about the validity of their date, they stuck fast to 2.6 million years. With a vested interest in the older date, Leakey too entrenched himself. But the evidence against him began to mount. Not only did the pig data between Omo and Turkana not correlate; neither did various other animal data including that for elephants, horses and antelopes. Matters were made worse when it became evident that some of the fieldwork undertaken by Leakey in the early years had been seriously flawed.

The KBS tuff controversy, as it was called, soon developed into a full-blown feud between two rival groups: Leakey's team in Kenya, adamant that the tuff was 2.6 million years old, and Clark Howell's American team in charge of the Omo River Valley sites, insisting that it was no older than 2 million years old. Both sides resorted to personal attacks. Howell's team possessed far more expertise and was scornful of Leakey's 'amateur' status and defective fieldwork. But it was Leakey's team that had made the most spectacular gains and achieved world-wide fame. A Rift Valley conference in London in 1975 was marred by scientists shouting at each other.

The turning point came when an independent test on new KBS tuff samples was carried out by Garniss Curtis at his potassium/argon laboratory at Berkeley, California. The oldest date he produced was 1.8 million years. Subsequent tests on the Koobi Fora pig fossils also confirmed they were no older than 2 million years, the same age as their counterparts at Omo and Olduvai.

Thus, the 1470 skull was assigned to an age contemporaneous with *Homo habilis*: about 2 million years old. Indeed, most scientists eventually concluded that it belonged to the same species, helping *Homo habilis* to gain recognition as a real biological entity.

Meanwhile, another challenger had emerged from the badlands of northern Ethiopia.

CHAPTER 8

HADAR

THE AFAR TRIANGLE is one of the most forbidding regions on earth, a tormented land of active volcanoes, blistering salt flats, boiling hot springs and sunken deserts lying far below sea level. During the day, it becomes a furnace, swept by scorching winds and sandstorms; during the night, its baking lava fields radiate immense stores of heat. Temperatures recorded in the Afar Triangle are among the highest in the world.

The local Afar nomads have an equally ferocious reputation. Several parties of European travellers venturing into their territory during the late nineteenth century were slaughtered and their bodies mutilated. A British mining engineer, Lewis Nesbitt, and two Italian companions became the first white strangers to survive an 800-mile-journey across Afar lands in 1928, enduring daily hardship and danger. In his account of the expedition, Nesbitt noted how Afar warriors wore the testicles of their victims as trophies to prove their prowess. In 1933, a young English adventurer, Wilfred Thesiger, set out from the Ethiopian capital, Addis Ababa, to explore the Awash River as it meandered through Afar territory, well aware of the natives' murderous reputation. 'A man's standing depended on the number of men he had killed and castrated', wrote Thesiger. British officials gave him a one-in-ten chance of surviving. But Thesiger found the Afar mostly hospitable. 'In time I learnt to tell at a glance, from the decorations he wore, how often a man had killed, just as I might tell from his campaign medals where a British soldier had served'.

In the post-war era, the Afar Triangle became of increasing interest to geologists pursuing new theories about continental drift and plate tectonics, ideas that were revolutionising the world of classical geology. Geologists discovered that the Afar Triangle is sited on a triple junction of three giant troughs in the earth's surface: the Red Sea, the Gulf of Aden and Africa's Rift Valley. These troughs were formed following the gradual separation of three major plates: the African (or Nubian), the Arabian and the East African (or Somali). Over a 30-million-year period, these three plates have been pulling apart—the African to the northwest; the Arabian to the northeast; and the Somali to the south. Sitting atop a cauldron of hot magma, the Afar junction is a geological 'hotspot' providing much of the energy behind these tectonic movements; on the surface, erupting volcanoes have spewed out a mass of lava, ash and rock. Over time, as the valley floor sank, the Afar region became a vast basin—a catchment area for rivers flowing down from the Ethiopian highlands, carrying silt, mud and sediment and forming lakes and floodplains.

In 1965, a thirty-year-old French geologist, Maurice Taieb, began exploring parts of the Awash Valley, an area that extended for about 30,000 square miles. Born in Tunisia, Taieb was studying for a doctorate at the University of Paris. He was familiar with desert environments and soon developed a particular fascination for the Afar region. Short of funds, he often continued to explore on foot, packing his supplies on donkeys and staying with Afar tribesmen along the way. During his third field season in 1969, he was driving across a gravel plateau near the northern bend of the Awash River when he came to an abrupt halt at the edge of a cliff.

Before him lay a stretch of the river fringed by dark green riverine forest; and beyond it an immense panorama of arid badlands. Once the bed of an ancient lake, the sedimentary terrain had been transformed by erosion into a landscape of crumpled hills and steep ravines. Streams from the adjacent highlands and searing wind had cut through 600 feet of prehistoric sediments, exposing multicoloured layers of rock and sand deposits. The site, Taieb related, was 'exceptional and fantastic'. An American geologist, Jon Kalb, who accompanied Taieb there two years later, described it as being like 'an enormous, flat-lying encyclopaedia of natural history with part of one page exposed on this hill, another in that ravine, another on the crest of a ridge'. The local Afar called the place 'Ahdi d'ar'. But Taieb recorded the name as Hadar. Exploring the area for a few brief hours before sunset, he found elephant, rhinoceros and pig fossils; the elephant teeth he picked up and took back to Paris were later estimated to be 3 million years old.

Taieb's discovery of Hadar and other promising sites in the Awash Valley brought international attention to the area. After meeting Taieb at the Pan-African Congress on Prehistory in Addis Ababa in December 1971, Louis Leakey agreed to help him with letters of recommendation. In 1972, Taieb returned to Hadar leading a group of French and American scientists on a detailed reconnaissance. A full-scale expedition—the International Afar Research Expedition—was launched the following year. Among its eighteen members was Yves Coppens, a thirty-three-year-old palaeoanthropologist from the Musée de l'Homme, who had worked in Chad before taking over as director of the French contingent of the Omo expedition.

The team also included a young American palaeoanthropologist, Donald Johanson. One year older than Richard Leakey, he possessed the same ruthless ambition and was to emerge as Leakey's chief rival. The only child of Swedish immigrants, born in Chicago in 1943, his interest in anthropology had first been kindled by reading Louis Leakey's account in *National Geographic* about Mary Leakey's discovery of *Zinj* at Olduvai in 1959. Louis, he recalled, was 'proof that a man could make a career out of digging up fossils'. As an undergraduate, he had attended a seminar that Louis had given at the University of Illinois and had fallen under his spell, as many others had done. 'Louis not only made the subject of paleoanthropology interesting', Johanson recalled, 'he made it come alive through his infectious imagination'.

As a graduate student, Johanson had spent three seasons with Clark Howell at Omo, where he gained a reputation for holding abrasive views about the French scientists working there. In his own account of the Omo project, he referred to the 'difficulties' of collaborating with the French. 'There was scarcely anything that the French did in the same way as the Americans', he wrote. 'Their attitude to work was less urgent'. According to an account of the Hadar expedition by Jon Kalb, Johanson was 'an unending source of divisiveness and tension in camp', frequently complaining about the work of Taieb and Coppens and indulging in malicious gossip. 'Johanson's opinions and complaints about people were oddly bitter and came in spurts, as though he were periodically seized with vitriol or some dire affliction'. He was obsessed, said Kalb, with finding hominids. 'He wanted to monopolise the expedition, to make the search for hominids its only purpose. He'd get upset if other work took precedence over his'.

Whatever his personal foibles, it was Johanson, half-way through the two-month expedition in 1973, who found the first hominid fossils. Exploring a gully near the base of the Hadar section, he idly kicked what looked like a hippo rib sticking out of the sand. It came loose, and he realised that it was probably the upper end of the shinbone of a small primate. As he was recording the spot in his notebook, he noticed another bone a few yards away—the lower end of a thighbone; next to it lay part of a knee joint. All three pieces fitted together perfectly, making a slight angle at the knee joint. At first Johanson thought the bones belonged to a monkey, but then he recalled that a monkey's thighbone and shinbone joined in a straight line. 'Almost against my will', he wrote, 'I began to picture in my mind the skeleton of a human being, and recall the outward slant from knee to thigh that was peculiar to upright walkers. I tried to refit the bones together to bring them into line. They would not go. It dawned on me that this was a hominid fossil'.

Desperate to confirm the hominid find, Johanson resorted to unorthodox methods. In his application for funds from the National Science Foundation to enable him to join the Afar expedition, he had hinted strongly at the possibility of finding hominid fossils, and yet, after a month in the field, with his money running out, he had so far nothing to show. 'What does a young man do on his first expedition, when he is given a two-year grant and has exhausted most of it the first year and has not found what he went out to look for?' he wrote. 'He wonders what he will do the second year. He wonders if he may not crash, if he may not get a reputation for irresponsibility before his career gets properly started. He sweats'.

The day after his discovery, Johanson set out from camp, accompanied by an American graduate student, heading for a nearby Afar burial mound. Reaching inside the mound, he retrieved a thighbone and took it back to camp to compare with the fossil. Other than size, they were virtually identical.

Johanson's find was highly significant. The age of the fossils based on associated fauna found at the same level was estimated to be 3 million years. If the find was confirmed to be hominid, it would provide the first evidence that human ancestors had walked upright as far back as 3 million years ago. In the Hadar camp, there was widespread celebration. Returning to Addis Ababa at the end of the two-month season, Johanson, unwilling to wait for proper authentication, held a solo press conference, making public his discovery, determined to gain full credit. Taieb, the expedition leader, who had by then flown back to Paris, read about it in the newspapers.

Back in the field a year later, while the case of the knee joint was still under discussion, Taieb's team made more dramatic finds. Within two weeks after setting up camp, Ethiopian researchers found parts of four hominid jawbones; one was a complete palate with all sixteen teeth in place. The bones seemed to combine both primitive and derived features—'peculiar blend of *Homo* and australopithecine traits, with a whiff of something more primitive', according to Johanson. Their age was estimated to be between 3 and 4 million years.

Well aware of the publicity value, the expedition's members agreed to announce their discovery to the media in Addis Ababa in a statement prepared by Johanson. Using the kind of rhetoric that Louis Leakey favoured, the statement proclaimed 'an unparalleled breakthrough in the search for the origin of man's evolution'. It declared: 'Discovery of the genus *Homo* older than 3.0 million years is a major step forward in our understanding of early man's evolution and represents a major revolution in all previous thinking concerning the origin of the group representing modern man'. Previous discoveries from Olduvai and Lake Turkana had only taken the origins of *Homo* back to slightly over 2 million years. 'We have in a matter of merely two days extended our knowledge of the genus *Homo* by nearly 1.5 million years. All previous theories of the origins of the lineage which leads to modern man must now be totally revised. We must throw out many existing theories and consider the possibility that man's origins go back to well over 4.0 million years'.

However overblown such claims seemed, Johanson's discovery the next month was truly spectacular. On the morning of 30 November 1974, exploring a fossil locality four miles from camp with an American graduate student, Tom Gray, Johanson spotted a fragment of an arm bone poking from the slope of a gully. At first sight, because of its small size, it appeared to be a monkey, but on closer inspection it lacked the characteristic flange of a monkey.

'My pulse was quickening', wrote Johanson in an article for *National Geographic*. 'Suddenly I found myself saying: "It's hominid"'. Higher up the slope, there were other fragments, and he began to realise that he might have found a skeleton. All around lay a multitude of bones: a nearly complete lower jaw, a thighbone, arm bones, ribs, vertebrae. 'The searing heat was forgotten. Tom and I yelled, hugged each other, danced, mad as any Englishmen in the midday sun'.

Back at camp, the celebrations continued long into the night. Amid the excited talk, a tape recording of the Beatles' song 'Lucy in the Sky with Diamonds'—a title originating from a drawing of a girl surrounded by stars—was played over and over again, at full volume. By morning, the skeleton itself was known as 'Lucy'.

Lucy was the most complete hominid specimen yet discovered. Excavations over the following three weeks yielded more bones; in all about 40 per cent of the main bones of an entire skeleton—discounting hand and foot bones—was recovered. The bones revealed that Lucy was indeed a young adult female hominid, capable of walking upright, standing at just over three feet tall, with an age estimated to be 3 million years (subsequently pinpointed as 3.18 million years). She possessed a mixture of features; some apelike, others humanlike; but in particular her brain appeared to be tiny, providing further confirmation that protohumans walked upright before their brains had begun to enlarge.

But whether Lucy belonged to the genus *Homo* or should be classified as an australopithecine remained uncertain. Introducing Lucy to the world's media at a press conference in Addis Ababa at the end of the 1974 season, Johanson was uncharacteristically circumspect: Lucy, he said, was either 'a small *Homo* or a small australopithecine'.

The discovery of Lucy propelled Johanson to worldwide fame. 'I was no longer an unknown anthropology graduate', he wrote in his autobiographical account, *Lucy*, 'but a promising young field worker with fossils dazzling enough to match those of paleoanthropology's certified supernova, Richard Leakey'.

But Johanson wanted more than to match Leakey; he was determined to surpass him and establish himself as pre-eminent in the field. So often did Johanson talk of his ambition that several of his colleagues thought he had become obsessed by the idea. Jon Kalb recalled that when gossiping about the Leakeys, Johanson had often been scathing about Richard Leakey's lack of academic credentials. The two fossil-hunters had nevertheless developed what appeared to be a firm friendship. After meeting Johanson at a conference in Nairobi in 1973, Leakey had gone out of his way to encourage him, making introductions, taking him into the field, offering advice. Both Richard and Mary Leakey had furnished letters commending Johanson's work in Hadar. The Leakey Foundation had come to his rescue with a $10,000 grant to enable him to get through the 1974 season there. But after finding Lucy, Johanson's ambition soared further. Maurice Taieb found his own leadership under challenge. 'He wants everything for himself', Taieb recalled, 'and it was all because he wanted to pass Richard'.

Johanson's run of luck continued to hold. When Johanson, Taieb and other members of the Afar Expedition reassembled in Hadar in 1975, accompanied by a camera crew, it took them only two weeks to make another remarkable discovery. A hillside known as site 333 yielded more than 200 pieces of bones and teeth, representing at least thirteen individuals—men, women and children—who were at first thought to have died together about 3 million years ago. They came to be known as the First Family—the earliest group of associated individuals ever found. (Subsequent research, however, indicated the bones may have been deposited in separate events, suggesting a more gradual accumulation.)

In an article for *National Geographic*, Johanson assigned them unequivocally to the genus *Homo*. He was ecstatic not just because the find was unprecedented, but because he had managed to trump the Leakeys. 'I've got you now, Richard!' he was recorded on camera as saying. 'I've got you now'.

But meanwhile, further Leakey finds were also receiving acclaim.

CHAPTER 9

LAETOLI

Tanzania

IN 1974, after completing a further six years of fieldwork at Olduvai, Mary Leakey decided to investigate a site called Laetoli some thirty miles from her camp. Named by the Maasai after a red lily which grows in profusion there, the Laetoli site had once consisted of flat savannah grassland, but numerous eruptions from a nearby volcano, Sadiman, twenty miles to the east, had built up layers of volcanic ash more than 400 feet thick.

Mary had visited Laetoli on three previous occasions (the first time with Louis Leakey in 1935, a second time in 1959, and more recently on a day's excursion in 1969). Exploring the area in 1974, Mary's team of fossil-hunters were soon rewarded; over a period of twelve days, they turned up fossils of thirteen hominids. Samples of lava from Laetoli were sent to Garniss Curtis in California for dating analysis. The results were spectacular. They showed that Laetoli's hominids ranged in age from 3.35 to 3.75 million years, making them the oldest hominid fossils ever found. One of the specimens—part of a lower jaw numbered LH-4—was destined to become the subject of an extraordinary furore four years later.

The fossils seemed to have a mixture of 'primitive' as well as distinctive *Homo* traits. But Mary insisted that they belonged to the genus *Homo* and not to *Australopithecus*. In her view, australopithecines were an offshoot of the hominid line, a contemporary of early man that had died out. But she was unsure where on the *Homo* line the Laetoli fossils should be placed. Hoping to find more and better specimens, she delayed putting a name to them. It was a decision that later she would come to regret bitterly.

In August 1975, Richard Leakey's team at Koobi Fora scored another success. At a location not far from the main camp, Bernard Ngeneo the Hominid Gang member who had found the 1470 skull spotted fragments of bone poking above the surface. After days of careful excavation, Ngeneo's find turned out to be a skull, with the face almost complete; but it resembled nothing else that had been found at East Turkana. While examining it, Leakey noticed strong similarities to specimens of *Homo erectus* from Zhoukoudian—Peking Man—thought to date back to between 500,000 and 700,000 years. Hitherto, few fragments of *Homo erectus* had been discovered in Africa. But Leakey was soon convinced of the skull's identity. 'There was no doubt', he wrote in his autobiography, 'that this was not *Australopithecus* nor even *Homo habilis* but rather *H. erectus*, a more immediate ancestor of ourselves'. Its brain size just reached the lower limits of *Homo erectus*. But what was particularly striking about it was its age. The initial estimate was between 1.3 million and 1.6 million years. But subsequent evaluation showed it to be 1.8 million years.

The discovery of skull 3733, as it was numbered, had ramifications that went far beyond the matter of its age. For it put an end to a theory about human evolution that had gained increasing attention in the 1960s and 1970s—the single species hypothesis. For several decades, most scientists had held fast to the notion that only one type of hominid could have been alive at any one point in time. When Robert Broom and John Robinson in 1949 announced the discovery of a novel species of hominid they called *Telanthropus*, suggesting that two extinct species had once coexisted, they set off a prolonged controversy. Louis Leakey faced the same furore in the 1960s by arguing that *Zinj* (a robust australopithecine) and *Homo habilis* had lived 'side by side'.

At the forefront of the single species theory in the 1960s were two influential American anthropologists, Loring Brace and Milford Wolpoff, of the University of Michigan. Like the architects of the modern synthesis of the late 1940s, they viewed human history as progressing by gradual change up a single evolutionary ladder that started with australopithecines and ended in modern humans. They conceded that there were measurable differences between individual fossils but saw no need to create new species to accommodate those differences. Their case relied on a piece of ecological theory—the principle of competitive exclusion—that maintained that two species with similar adaptations could not occupy the same ecological niche in the same place at the same time; sooner or later, one of the two species would out-compete the other. Applying the principle of competitive exclusion to hominid evolution, they argued that because of cultural adaptation—such as toolmaking—only one species could have existed at any given period.

The discovery of 3733—a clear example of *Homo erectus*—blew this theory apart because it had been found at the same stratigraphic level as a skull numbered 406 (that Richard Leakey had discovered in 1969) and that had proved to be a clear example of *Australopithecus boisei*. The differences between them were easily recognised. The British anatomist Alan Walker described 406 as 'an eating machine'—it possessed enormous teeth, jaws and chewing muscles that were anchored to striking bony crests at the top and back of the skull and to its strong cheekbones. By contrast, 3733 was 'a thinking machine'—its cranium was dominated by a large, globular braincase housing a brain that was almost twice as big as that of 406. Yet both had lived at the same 'time horizon'. In a paper for *Nature*, Leakey and Walker wrote: 'The new data show that the simplest hypothesis concerning early human evolution is incorrect and that more complex models must be devised'. An article in the American magazine *Newsweek*, discussing the recent spate of finds from Koobi Fora and Hadar, described Leakey's 3733 skull as 'the most impressive' of all.

In December 1975, at the end of the Hadar season, Johanson took his First Family to Nairobi to show the Leakeys and to gain their opinions. 'When I spread out the haul of new bones, they were an instant sensation', Johanson wrote. 'Nothing that combined their extreme antiquity, their remarkable quality and their profusion had ever been encountered before'. Examining the fossils, both Mary and Richard Leakey concurred with Johanson that they appeared to be early examples of the genus *Homo*.

Among the scientists who had gathered around a specimen table at the Nairobi Museum was Tim White, a twenty-five-year-old graduate student from the University of Michigan who had spent two seasons with Richard Leakey's team at Koobi Fora. Although known for his prickly and abrasive temperament, White had impressed Leakey with his sharp intellect and passion for scientific accuracy. When Mary Leakey was seeking specialist help with the anatomical description of her Laetoli hominids, Richard had recommended White. Mary Leakey, too, had been impressed by the quality of White's work. But on this occasion White was unusually reticent.

Johanson noticed him 'lurking' in the background—'an owl-eyed young man with thick glasses, lank blond hair and a white lab coat'—and assumed him to be shy. But White was simply wary of Johanson's brash presentation. 'I was just being sensibly cautious', White subsequently recounted in conversation with Johanson. 'There you were, the smooth young hotshot, shooting off your mouth about all your great fossils. I'd never met you before. I didn't know if you could tell a hippo rib from a rhino tail. I was just waiting for you to fall on your face, say something really dumb'.

In due course, White stepped forward to examine the Hadar specimens. 'I think your fossils from Hadar and Mary's fossils from Laetoli may be the same', he said.

It was an idea that would eventually explode into yet another controversy.

The 1976 season at Laetoli brought further stunning success. By sheer chance, a group of scientists on a short visit from Kenya stumbled across an entirely new phenomenon: ancient footprints. Walking back to Mary's camp one morning in July, they began to amuse themselves by tossing lumps of dried elephant dung at each other. Searching for more dung supplies, Andrew Hill, a palaeontologist who worked for the Nairobi Museum, jumped down into a flat gully and spotted what appeared to be elephant tracks. Falling to his knees to examine them, he realised they were not fresh tracks but fossilised elephant footprints preserved in a layer of volcanic ash. Scattered around them lay the prints of other animals—antelope, buffalo and giraffe. Tiny indentations in the tuff later turned out to be caused by ancient raindrops.

The focus of attention at Laetoli switched to this new locality—Site A, as it was called. The profusion of prints uncovered was extraordinary. Site A alone contained 18,400 prints, ranging from elephant to insect tracks. But Mary's team soon discovered that the same layer of volcanic ash—the Footprint Tuff—occurred elsewhere at Laetoli. By the end of the 1976 season, they had found three more sites with animal prints, and the eventual total reached eighteen.

Piecing together what had happened at Laetoli, an American geologist, Richard Hay, concluded that about 3.6 million years ago, the volcano Sadiman began a spate of eruptions, showering the plains below with light ashfalls. The ashfalls continued for a period of several weeks, leaving a series of fine laminations on the ground, just as the dry season was giving way to the onset of rains. The ash contained a high concentration of the mineral carbonatite, which, when mixed with rain, formed a soft cementlike surface that retained the prints of animals as they moved across it. When the surface dried out in the sun, it became rock hard, preserving the prints. A few weeks later, a final eruption from Sadiman sealed the Footprint Tuff with a thick layer of ash. Over time, that layer had eroded.

With so much evidence of animal prints to hand, Mary's team began to keep a lookout for any sign of hominid prints. In September, researchers spotted what looked like a hominid trail at Site A—four prints bearing the characteristic trace of a humanlike big toe. Initially, Mary remained sceptical, but after further study during the 1977 season, she concluded that they were indeed hominid. In February 1978, she announced her finding to the world at a Washington press conference. Several experts expressed doubts at the time, but later in the year, as the hunt for hominid prints continued, Mary Leakey was to achieve a spectacular success.

At his laboratory at the Cleveland Museum of Natural History in Ohio, Donald Johanson began the task of analysing the remarkable collection of Hadar bones he had brought back from Ethiopia. The haul included Lucy and the First Family and, in all, amounted to 350 separate fossil pieces from males, females and juveniles dating back 3 million years.

To help him with the analysis, Johanson invited Tim White to join him in Cleveland. Knowing that White was in possession of casts from the Laetoli fossils that he had been describing for Mary Leakey, Johanson asked White to bring them with him so that a comparison could be made with the Hadar fossils.

When making his first assessment of the Hadar fossils, Johanson had assigned Lucy as an australopithecine and the First Family as *Homo*. Lucy was different from the others, he had decided; she was much smaller, more primitive, with only a tiny brain. The notion that the First Family fossils belonged to *Homo* had been reinforced in 1976 when archaeologists working in a gully three miles from the main camp in Hadar discovered stone tools that were dated as 2.5 million years old—the oldest known stone tools in the world. Johanson assumed that only *Homo* could have made them.

In Cleveland, White argued that Lucy and the First Family belonged to the same species. The reason for the difference in their size, he suggested, was sexual dimorphism: females were smaller than males of the same species. What both noticed was that the Hadar fossils and the Laetoli fossils, though taken from sites a thousand miles apart and separated in age by more than half a million years, seemed decidedly similar.

By the end of 1977, Johanson and White had reached an agreed position. They concluded that there was compelling evidence to place not only Lucy and the First Family within the same species but Mary Leakey's Laetoli fossils as well. Although markedly different in size, they were morphologically alike. At one end was Lucy: little more than three feet tall, probably weighing about sixty pounds; at the other end were individuals five feet tall, weighing up to 150 pounds. But all were said to be members of a single bipedal species.

Johanson and White also decided to classify the Hadar-Laetoli fossils as belonging to *Australopithecus* rather than *Homo* because they lacked an enlarged brain—the hallmark of the genus *Homo*—while possessing many characteristics common to australopithecines. But because several features—such as the more primitive teeth—differed from other known australopithecines, they believed it necessary to create a new species that they called *Australopithecus afarensis*.

As both Johanson and White knew, all this was bound to provoke controversy. No new hominid species had been created since Louis Leakey had named *Homo habilis* thirteen years before, triggering uproar. But the two Americans decided to go even further: to rearrange the entire family tree. The most widely accepted family tree in the 1970s showed *Australopithecus africanus* as ancestor both to *Australopithecus robustus* and to *Homo habilis,* a branch that led to *Homo erectus* and *Homo sapiens*. But the position that Johanson and White now adopted was that *Australopithecus afarensis* was the common ancestor to later australopithecines and to *Homo*, spanning a period between 4 and 3 million years ago. *Australopithecus africanus* was relegated to an intermediate stage in the doomed australopithecine line. At about 3 million years, according to Johanson and White, the human line began to emerge. By 2 million years, it had become recognisable as *Homo*. For about a million years, *Homo* and australopithecines had lived side by side. By 1 million years ago, there were no australopithecines left. They had all become extinct.

Needing a type specimen for *Australopithecus afarensis*—the reference model for the species by which all others must be judged—Johanson and White chose a fossil not from the Hadar collection but from one of Mary Leakey's much smaller group of fossils—part of a lower jaw known as LH-4 (Laetoli hominid 4). With Mary Leakey's approval, White had previously published a meticulous description of the Laetoli fossils, leaving out any comment about their affinities. But now LH-4 was assigned as the 'name-bearer' of the new species. The advantage to Johanson and White was that it enabled them to claim an older date for *Australopithecus afarensis*—half a million years older—than otherwise would have been the case if they had confined their analysis to the Hadar fossils; the Laetoli fossils had already gained fame as the world's oldest hominids. The choice of LH-4 also meant that Mary Leakey's Laetoli fossils—which she had not yet named—would henceforth become known as *Australopithecus afarensis.*

Johanson assumed that Mary Leakey would be pleased with this arrangement and hoped to be able to persuade her to lend her name to an article that he planned to write in conjunction with Tim White and Yves Coppens, the French co-leader of the Afar Expedition. From the outset, however, Mary Leakey made clear her objection in the first place to the term *Australopithecus*. In answer to a letter from Johanson in December 1977, she wrote: 'I do not think *Australopithecus* is correct. It is a lousy term, based on a juvenile [the Taung child from South Africa] for which there is doubt as to whether it is *A. africanus* or *A. robustus*. Nor is it a direct ancestor of *Homo*, as all of us people agree'.

During a lecture tour of the United States in February 1978, Mary met White at his office in the anthropology department at Berkeley, once again stating her objection to the use of the term *Australopithecus*. She also spoke on the phone to Johanson. 'I said I didn't think the Laetoli specimens were *Australopithecus*', she recalled. 'I objected to the term. That has been consistently my view'.

Johanson and White nevertheless proceeded to draft a 'naming' paper for publication, describing *Australopithecus afarensis* as a new species and using LH-4 as the type specimen. Johanson also made plans to make a public announcement.

The occasion he chose was a Nobel Symposium in Stockholm in May 1978, a six-day event organised by the Royal Swedish Academy of Sciences as part of national celebrations commemorating the 200th anniversary of the death of the Swedish botanist Carolus Linnaeus. As the son of Swedish emigrants, Johanson felt particular pride in being invited to speak at such a grand affair.

It was an event to which Mary Leakey had also been invited as a special guest of honour. She was due to be presented with the Golden

Linnaen Medal by Sweden's King Gustav in acknowledgment of her contributions to biological sciences—the first woman ever to receive the award.

During the scientific discussions, it fell to Johanson to deliver his lecture in advance of Mary Leakey. While Mary sat listening in the audience, much to her fury Johanson dwelt not just on the Hadar discoveries but on the Laetoli fossils that were the result of her own work and about which she was due to speak next after a coffee break. At least half of his lecture was devoted to Laetoli. During the coffee break, she vented her exasperation to Richard Leakey: 'Did you hear that?' she demanded. 'That fellow talked about my fossils. He talked about my site. How am I going to give my paper now? It's all been said'. Ten years later, in conversation with Virginia Morell, she was still exasperated by what had happened. 'All the things I was going to say about my site that I had been funded for, Don talked about as if they were his own', she said. 'It was for me to announce those things; not him . . . It was so impertinent'.

To Johanson's surprise (and disappointment) his announcement about *Australopithecus afarensis* stirred little interest in the audience. No one asked a question or ventured an opinion. But for Mary, it marked the beginning of an open rift. When she subsequently learned that Johanson had included her as one of the joint authors of the naming paper on *Australopithecus afarensis* sent for publication to *Kirtlandia*, the house journal of the Cleveland Museum, she insisted that her name be removed. Publication was delayed and the paper had to be reprinted.

Relieved to be back in the field at Laetoli, Mary Leakey was rewarded with one of the most remarkable discoveries in the history of palaeoanthropology. In July 1978, a member of her team, Paul Abell, spotted what looked like the heel part of a hominid print on a section of hardened ash one mile from the camp. Initial excavations revealed two distinct human footprints. By the end of the 1978 field season, the Laetoli team had uncovered a total of forty-seven prints belonging to three individuals walking together on a trail stretching over seventy-five feet. Another twelve feet of the trail were uncovered during the 1979 season.

What the footprints proved—indisputably—was that 3.6 million years ago, human ancestors had walked upright with a free-striding gait and that the shape of their feet was strikingly similar to that of modern humans.

While Mary Leakey was savouring this triumph, the world of anthropology was engulfed in controversy over the advent of *Australopithecus afarensis*.

CHAPTER 10

Bones of Contention

WHAT BECAME known as 'The Battle of the Bones' began in January 1979 when an article by Johanson and White explaining their new interpretation of the origins of humankind was published as the main feature in the American journal *Science*, prompting widespread media interest. The press reported that 'a previously unknown human ancestor' combining a small-brained apelike head with a fully erect body had been discovered in Africa. In a front-page article, the *New York Times* pointed out that the new species—*Australopithecus afarensis*—presented a major challenge to conventional theories about human evolution.

Among the anthropological community, opinion about the merits of the Johanson-White hypothesis was divided. Attacks came from several different quarters. Richard Leakey rejected the notion that *Australopithecus afarensis* was the common ancestor of all later australopithecines and *Homo*; the common ancestor, he said, had not yet been found. He was ready to accept that Lucy was a new species of australopithecine; but he maintained that among the other Hadar fossils were examples of two different populations: *Homo* and *Australopithecus*.

And he held fast to his view that the *Homo* lineage had much deeper origins than the 2-million-year threshold proposed by Johanson-White.

Mary Leakey's response was far sharper. Furious at the way Johanson and White had purloined the Laetoli fossils for their own use, she dismissed their work at a Washington press conference in March 1979 as 'not very scientific'. She believed that their real purpose was not so much to establish a link between the Hadar and Laetoli fossils as to secure for *afarensis* the older, proven date of the Laetoli fossils. The obvious choice for a type specimen for *afarensis*, she pointed out, was not LH-4 from Laetoli but Lucy. 'It is regrettable', she told Roger Lewin, 'that the type-specimen selected should be a worn mandible from Laetoli, when much better-preserved specimens are available from Afar itself'.

Other scientists lined up to support her. The American geologist Richard Hay described the use that Johanson-White had made of the Laetoli fossils as 'a form of scientific theft'. The eminent zoologist Ernst Mayr, an expert on taxonomy, was scathing about the way they had taken the type specimen from Tanzania but the name from Ethiopia. 'If you select a geographical locality for the name, then you have no choice but to select the type-specimen from the same locality'. The two sets of fossils were located 1,000 miles apart and separated by half a million years in time, leaving open the possibility that they were geographic variants induced by climatic and other environmental conditions. 'Every species consists of numerous local populations differing to the degree of their isolation'. The proper decision therefore, said Mayr, would have been to name Lucy as the type-specimen. 'It was a horrible thing that Johanson did', he told Virginia Morell, ' . . . and made everyone's hair stand on end'. He wanted the International Commission on Zoological Nomenclature to suppress Johanson's type designation and designate instead a specimen from Afar.

Further doubt about the Johanson-White hypothesis was cast by Yves Coppens, the French co-leader of the Afar expedition who had previously agreed to be listed as a joint author in the naming paper published by *Kirtlandia*. Coppens published an article saying that, like Richard Leakey, he recognised at least two species among the Hadar fossils, not just one. As well as bones belonging to *afarensis*, he detected a primitive species of *Homo* present, too.

From South Africa, Phillip Tobias, Professor of Anatomy at the University of the Witwatersrand, weighed in with his own broadside. Tobias saw no reason *Australopithecus africanus* should be usurped by *afarensis* as the progenitor of all later hominids. The Laetoli and Hadar hominids, he declared, were indeed subspecies; but of *africanus*, not *afarensis*. He suggested that the Hadar hominids should be named *Australopithecus africanus aethiopicus* and the Laetoli hominids *Australopithecus africanus tanzaniensis*. Addressing an international scientific meeting in London in March 1980, he called for the term *afarensis* to be scrapped: 'Since the tying of the name "*A. afarensis*" to the Laetoli fossils is manifestly inappropriate and since it is considered that the case for "*A. afarensis*" has not been established, it is proposed formally that the name "*A. afarensis*" be suppressed'.

Despite all the hubbub, the Johanson-White hypothesis gradually gained favour, and *Australopithecus afarensis* duly became accepted by most of the scientific community as the oldest member in the pantheon of ancient ancestors yet discovered. The picture that emerged was of a hominid weighing from seventy-five to 125 pounds, between three and four feet tall, with a brain volume of between 400 and 500 cubic centimetres, only a little larger than the average brain size of a chimpanzee.

After studying the Hadar hominids, Owen Lovejoy, an expert on the biomechanics of locomotion at Kent State University in Ohio, concluded that *afarensis* hominids were well-adapted bipeds. 'They look incredibly primitive above the neck and incredibly modern below. The knee looks very much like a modern human joint; the pelvis is fully adapted for upright walking; and the foot, although a curious mixture of ancient and modern, is adequately structured for bipedalism. Some of the bones in the feet are slightly curved, and look rather like the bones you'd expect to see in its ancestor who climbed trees. But I believe that the curvature in the *afarensis* foot bones is well suited for walking on soft, sandy terrain; it probably inherited the curved feet from its tree-climbing ancestors, but the shape has been made use of in a different way'.

The upright gait of *afarensis* led many palaeoanthropologists to conclude that Lucy and her kind, though descended from tree-dwelling ancestors, were terrestrial creatures that spent their whole time at ground level. But further studies by researchers at the State University of New York at Stony Brook presented a different picture. They agreed that while on the ground, Lucy had clearly functioned as a biped. But they noted that Lucy possessed relatively long arms, ape-like shoulder joints, powerful wrists and curved toes, all suggesting that *afarensis* retained a tree-climbing capacity and still spent a good deal of time in the trees.

All this required some new thinking about human evolution. The common view at the time of Lucy's discovery was that upright walking had evolved from the need by proto-humans to free their hands to make tools and to carry them and other objects around. But *afarensis* showed that hominids walked upright as much as a million years before stone tools appeared in the archaeological record. A new theory was needed, therefore, to explain why human ancestors walked upright in the first place.

While palaeoanthropologists were preoccupied with the business of fossils, other scientists began to make increasing inroads into what they regarded as their domain, provoking a new set of arguments. At the forefront was a group of biochemists and molecular biologists who argued that molecular evidence was more reliable than morphological evidence in uncovering evolutionary histories. In 1962, Emile Zuckerkandl and Linus Pauling, two pioneering scientists at the California Institute of Technology, used the term 'molecular anthropology' to describe this new field of research. Their work showed that the structure of molecules of blood proteins—specifically haemoglobin—changed over time with such regularity that it provided a molecular 'clock' that could be used to help construct evolutionary, or phylogenetic, trees. In a paper published in 1965, they referred to molecules as 'documents of evolutionary history'.

They were followed by Morris Goodman, a biochemist at Wayne State University's School of Medicine in Detroit, who produced a phylogenetic tree based on immunological data that challenged accepted notions about the evolutionary relationship between great apes and humans. Hitherto, zoologists had classified the great apes under one family, the Pongidae, while placing humans in their own family, the Hominidae. Goodman's tests on blood proteins showed that African apes—chimpanzees and gorillas—were more closely related to humans than were Asian apes—orang-utans and gibbons. And he proposed that because of their genetic propinquity, humans, chimpanzees and gorillas should be placed in the same family. Furthermore, his tree showed that humans, gorillas and chimpanzees had split from a common ancestor at a similar period of time; the conventional view was that humans had separated at a much earlier time.

In the late 1960s, two Berkeley biochemists, Vincent Sarich and Allan Wilson, set out to put dates on the branching points in Goodman's phylogenetic tree, applying the molecular clock theory developed by Zuckerkandl and Pauling. The prevailing wisdom among palaeoanthropologists, based on fossil evidence, was that the divergence between apes and the human line had occurred between 30 and 15 million years ago. The date that Sarich and Wilson came up with in 1967 was far more recent: 5 million years ago.

A running feud broke out between palaeoanthropologists and the molecular school. Outraged that molecular scientists should impinge on their own territory, palaeoanthropologists challenged the reliability of the molecular clock theory. The fossil record, they insisted, was a far more accurate measure. Addressing a meeting at the New York Academy of Sciences in 1968, John Buettner-Janusch of Duke University poured scorn on the molecular approach. 'I object to careless and thoughtless statements about evolutionary processes in some of the conclusions drawn from the immunological data mentioned', he said. 'Unfortunately there is a growing tendency which I would like to suppress if possible to view the molecular approach to primate evolutionary studies as a kind of instant phylogeny. No hard work, no tough intellectual arguments. No fuss, no muss, no dishpan hands. Just throw some proteins into the laboratory apparatus, shake them up, and bingo!—we have an answer to questions that have puzzled us for at least three generations'.

But further research by Sarich and Wilson, involving DNA molecules as well as proteins, confirmed the validity of their original approach. 'One no longer has the option of considering a fossil specimen older than about eight million years a hominid, no matter what it looks like', proclaimed Sarich in 1971.

For more than a decade, many palaeoanthropologists fought a rearguard action against the molecular school but steadily lost ground. By the early 1980s, a new family tree using molecular techniques had gained widespread acceptance. It showed the ancestor of gibbons splitting off first from the line leading to humans more than 20 million years ago; followed by the ancestors of orang-utans splitting off at 16 million years ago; by gorillas at about 10 million years ago; by chim-

-panzees at between 6 and 7 million years ago; and finally, by australopithecines splitting off from the line leading to early *Homo* at between 3 and 4 million years ago.

While the molecular revolution was gathering momentum, new schools of thought were emerging about the process of evolutionary change. Since the 1950s, the ruling doctrine to which most scientists subscribed was the evolutionary, or modern, synthesis, the theory that viewed evolution as the steady accumulation of small genetic changes over long periods of time under the guiding hand of natural selection. Its defining feature was gradualism. New species were the result of gradual change involving a whole population.

But in 1972, two American invertebrate palaeontologists, Niles Eldredge of the American Museum of Natural History and Harvard University's Stephen Jay Gould, presented a counter-theory. From his studies of trilobites—fossil marine arthropods dwelling on the sea bottom—Eldredge had noticed a distinct lack of evolutionary change; for millions of years, one type of trilobite had remained unaltered until environmental change had enabled a new species to invade and replace it. Gould had observed the same kind of pattern from his studies of an ancient species of land snail. It appeared to both of them that evolutionary change occurred not so much as part of a gradual process but in relatively short, sporadic episodes, with most change being concentrated in branching events in a geographically restricted subset of a population. Once evolved, new species, with their own peculiar adaptations, behaviours and genetic systems remained unchanged for long periods of time, often for several million years. In a paper published in 1972, they called this process 'punctuated equilibrium' and posed it as an alternative to the doctrine of 'phyletic gradualism'.

They claimed that their theory provided a more convincing explanation of the gaps that seemed to occur in the fossil record than the one given hitherto by the 'gradualist' school. They pointed out that if evolution had occurred as a result of slow, gradual and continuous change over the generations (as the gradualists maintained) then the fossil record should show it. But the record appeared to contain gaps. In the past, these gaps had been attributed to the difficulty that researchers faced in finding enough fossils to fill in the record. But even though a plethora of fossils had since been discovered, the 'gaps' were still said to be there. The explanation, said Eldredge and Gould, was that evolution occurred in bursts of change followed by long periods of stasis, or non-change. What the fossil record showed, they argued, was 'breaks' in the pattern of evolution, not missing links; in some cases, changes may have occurred so rapidly that intermediate forms had not been preserved in the fossil record.

They also contested the gradualist assertion that human evolution had occurred as a straight line of continuous transformation of one species into the next. They argued instead that human evolution resembled a multibranched bush of diversity, with many species coming into existence and almost as many dying out.

The punctuated-equilibrium theory sparked an intense debate that lasted for decades. What everyone agreed, however, was that human evolution was a far more complex business than had once been thought.

CHAPTER 11

Turkana Boy

While working at Koobi Fora, Richard Leakey often gazed across the jade-green waters of Lake Turkana, wondering what secrets lay on its western shore. In 1980, he despatched Kamoya Kimeu and the Hominid Gang to begin a series of preliminary explorations there.

Since joining the Leakeys at Olduvai in 1960 as an apprentice fossil-hunter at the age of twenty, Kimeu had led the Hominid Gang on a series of expeditions, scouring sites at the Omo River Valley, Koobi Fora and Laetoli. His skill as a fossil-hunter had become legendary. His team had found more primate and hominid fossils than any other group of professional palaeoanthropologists anywhere. Upon seeing Lucy for the first time, Kimeu told Don Johanson: 'If you found that, think what I could find'[1]

The key to Kimeu's success according to the British anatomist Alan Walker, a friend and colleague was his perseverance. 'He walks the same territory over and over again, changing courses around obstacles, and he tells his people to do the same', Walker wrote in *The Wisdom of the Bones*. 'If you walked to the left around this bush yesterday, then walk to the right today. If you walked into the sun yesterday, then walk with the sun at your back today. And most of all, walk, and walk, and walk, and *look* while you are doing it. Don't daydream; don't scan the horizon for shade; ignore the burning sun even when the temperature reaches 135 degrees F. Keep your eyes on the ground searching for that elusive sliver of bone or gleaming tooth'.

Kimeu also possessed a hawklike talent for spotting shapes on the ground that others would miss. And it was his expertise that would lead to yet another spectacular discovery.

After setting up camp in August 1984 in a grove of acacia trees on the banks of a sand river called Nariokotome (three miles inland from the lake) Kimeu's team spent two weeks exploring the area, finding an array of animal bones but no hominid fossils. Frustrated by the lack of results, Kimeu decided to move to another area after a day of rest. While other members of the team spent the day relaxing, Kimeu set off once more, choosing to look at a small hill on the opposite bank of the dry river bed, some 300 yards from camp.

To the untrained eye, it seemed an ordinary place: a scattering of black lava pebbles on the slope; a goat track snaking past a ragged thorn tree; a large salvadora tree where local Turkana children would gather to eat its pungent sweet-sour berries. But among the pebbles and dried leaves and sticks on the ground, Kimeu spotted a fragment of bone; it was no bigger than two postage stamps, one inch by two inches. Picking it up, he recognised it as belonging to the cranial vault of a hominid—the bony covering of the brain.

In the days that followed, as excavations got underway, more and more bones were discovered: parts of the skull, shoulder blades, ribs and pelvis. After four weeks of digging and sieving, Richard Leakey's Turkana team managed to retrieve enough body parts to assemble a skeleton. What was more, it was a skeleton of *Homo erectus*—the first one that anyone had ever seen.

It had been nearly 100 years since the Dutch physician Eugène Dubois had come across the first evidence of *Homo erectus (Pithecanthropus)* in Java. Subsequent discoveries in Asia of *Homo erectus* bones

pp. xxi–xxii

during the 1920s and 1930s had lent weight to the notion that the origins of humankind were to be found there rather than in Africa. Estimates of the date for these Asian fossils ranged from 700,000 to 500,000 years. During the 1960s, fragments of *Homo erectus* had been uncovered from a variety of sites at Olduvai with dates ranging from 1 million to 500,000 years. Then in 1975 came the discovery of the skull 3733 in East Turkana thought, initially, to date back to about 1.5 million years. Yet, for all the bits and pieces of *Homo erectus* that had been found, much remained unknown.

The Turkana skeleton gave a far more complete picture. Aside from hand and foot bones, nearly an entire skeleton was recovered. The teeth revealed it to be a youth, about nine years old; the pelvis suggested it was a male; the leg bones confirmed it walked upright. It possessed a full forehead and a round, smooth cranium, indicating a brain size significantly larger than an australopithecine; the brain size was later calculated to have been about 830 cubic centimetres, or 880 cubic centimetres for a fully grown adult, about two-thirds the size of a modern man's brain. Its height was estimated to be about five feet three inches—as an adult it would perhaps have reached six feet. In general, its body proportions matched those of modern people.

Of crucial importance was the date when it lived. Most of its bones were found lying right on top of a layer of volcanic tuff which the geologist Frank Brown assessed to be about 1.65 million years old. The age assumed for Turkana Boy, therefore, was 1.6 million years.

After an exhaustive study of the skeleton, Alan Walker concluded that Turkana Boy and his species *Homo erectus* represented an impressive advance from their predecessors. Long-legged and immensely strong, they used techniques to make stone tools that were more developed than before. They were also successful in obtaining higher-quality foods, almost certainly by hunting. Although they were not capable of speech, they lived in groups with strong social ties. Yet for all these physical attributes, *Homo erectus* was still essentially an animal—'a clever one, a large one, a successful one—but an animal nevertheless'. Turkana Boy, wrote Walker, may have looked much like a human, but he almost certainly did not act like a human. 'There was no human consciousness within that human body'.

One year later, during a second season at West Turkana, Alan Walker alighted upon another remarkable fossil. While searching a site on the Lomekwi River, twenty miles south of Nariokotome, Walker came across a small pile of stones left by a member of the Hominid Gang a few weeks before to mark the location of a piece of dark-coloured fossil. Walker picked up the fossil, examined it, then put it down. Then he picked it up again. It was part of an upper jaw with enormous tooth roots. For a moment he thought it was from some kind of extinct antelope. Then he saw another piece of bone that looked like the front of the skull of a large monkey. When he turned it over, he realised from its frontal sinus that it was hominid. With increasing excitement, excavations began.

The 'Black Skull', as it came to be known because of its dark patina, caused the family tree to be redrawn once more. It belonged to a robust australopithecine, a hominid similar to Olduvai's *Zinj*—*Australopithecus boisei*—but with more pronounced features: an ape-like jaw, massive brow ridges and a huge sagittal crest running the length of the brain case. The brain was extremely small for a hominid, no more than 410 cubic centimetres. Walker and Leakey decided to describe it as a 'hyper-robust' example of *Australopithecus boisei*. But whereas *Zinj* had been dated as just under 2 million years old, the Black Skull, according to the age of deposits where it was found, was estimated to be 2.5 million years old, predating almost everything else that had been thought of as a robust type of australopithecine. It was acclaimed in *Time* magazine as 'the most exciting find since Lucy'.

The ramifications of discovering a 2.5-million-year-old robust australopithecine were considerable. The family tree hitherto accepted by most palaeoanthropologists showed the doomed australopithecine line running from the smaller-toothed *africanus* to the big-toothed *robustus* to *boisei* before reaching a dead end about 1 million years ago. When drawing up their family tree in 1979, Johanson and White had followed the same pattern, while placing *afarensis* as the common ancestor of the australopithecine line as well as the *Homo* line. But the appearance of a big-toothed australopithecine at such an early stage in the australopithecine line suggested that something was wrong with this phylogeny. 'Whichever way you look at it', Dr Fred Grine, of the State University of New York at Stony Brook, wrote in *Science* magazine, 'it's back to the drawing board'.

There were other ramifications. Richard Leakey maintained that the discovery of the Black Skull cast doubt on the pivotal role of *afarensis* that Johanson claimed for it. On previous expeditions to East Turkana, Leakey's team had found evidence of two kinds of hominid living side by side: the 1470 skull, a *Homo habilis* specimen dated to about 2 million years; and a gracile australopithecine, as yet unnamed. The addition of a robust australopithecine in West Turkana meant that at least three types of hominid had shared the Turkana Basin between 2.5 million and 2 million years ago. In those circumstances, Leakey argued, it was hardly possible that all three hominids had sprouted from the single stem of a 3-million-year-old *afarensis*, as Johanson had insisted. 'This throws cold water on the notion that as recently as three million years ago there was only one species [of early human] which gave rise to the others', Leakey told reporters in Nairobi.

The feud between the Leakeys and Don Johanson had become even more intense in the 1980s. In his account of discovering Lucy, published in 1980, Johanson, relying on hearsay and gossip, had made various disparaging remarks about the Leakeys. Richard Leakey described his book as 'a cheap, journalistic slap at Mary and me'. Mary had been outraged by the 'lies' she claimed Johanson had written about her and her work, all the more so because of the help and hospitality she had offered him over the years. Among the errors to which the Leakeys pointed was a remark supposed to have been made by Louis Leakey when he first saw *Zinj*: 'It's nothing but a *goddamned* australopithecine', Johanson reported him as saying, a word he had never been known to use. A television show in New York in 1981 in which Johanson and Richard Leakey sparred in public further exacerbated matters.

Leakey's exploits in West Turkana now thrust him once more into the limelight. His team there had found not only Turkana Boy and the Black Skull but also remains of three types of Miocene apes, 17 million years old, two of which were new to science. In 1984, the *New York Times* devoted nearly an entire page to an article praising the Leakey family. Their 'towering reputation', it said, was well deserved.

Johanson, by contrast, had endured a lean period. He had not made a fossil discovery since 1977. His hopes of resuming his expedition to Hadar had been thwarted by a moratorium imposed by the Ethiopian authorities in 1982 on foreigners wanting to undertake palaeoanthropological research. One of the reasons for the ban had been Johanson's admission in his book *Lucy* of his grave-robbing exploit in Hadar. When Johanson subsequently asked Leakey for permission to study new fossils in Kenya, Leakey had rebuffed him. 'I consider you a scoundrel', Leakey told him by letter.

Johanson's next course of action sealed the rift. Banned from Ethiopia, shunned in Kenya, Johanson set his sights on Olduvai, hitherto regarded by the Leakeys as their backyard. Although Mary had retired from active fieldwork there in 1983 and had moved back to Nairobi, her lifetime's work at Olduvai had given her an abiding interest in what happened there.

Without informing Mary, Johanson applied for permission from the Tanzanian authorities to carry out research at Olduvai. He only told

Mary of his intentions shortly before leaving for Olduvai on a preliminary survey in 1985. Mary (according to Richard Leakey) found the news 'terribly upsetting'. But Johanson was unrepentant. 'In my mind', he wrote in *Lucy's Child*, 'it was time to stop thinking of the Gorge as the Leakey's living room. It deserved a future as well as a past'.

In July 1986, Johanson returned to Olduvai at the head of a field expedition, 'dropping my bag in what had been Mary Leakey's hut'. Among his team was Tim White, who had first suggested that the Hadar and Laetoli fossils were from the same species. They were soon rewarded with success. While walking across the bottom of the gorge, only three days after setting up camp, White spotted a fragment of a hominid elbow. Alongside him, Johanson saw the upper part of an arm bone. 'The Leakeys took thirty years to find a hominid at Olduvai', White gloated. 'We've got one in three days'.

In the five weeks of excavation that followed, some 300 fossil fragments were recovered. Many were unidentifiable. But from the remainder, Johanson's team managed to piece together parts of a skull, a palate with some worn teeth and small segments of both arms and legs, enough for Johanson to be able to claim 'a partial skeleton'. What emerged, according to Johanson and White, was a creature smaller than Lucy, standing about three feet three inches, but with longer arms in relation to its legs. Because it was so small, they deduced that it was female. Although it was as apelike as Lucy, it was dated as being a million years more recent in time—about 1.8 million years.

After studying the lower jawbone and teeth, Johanson and White concluded that this creature—Lucy's child, as it came to be known—was a specimen of *Homo habilis*. 'We had found the first skeleton known of the earliest species of *Homo*', Johanson claimed.

But naming Lucy's child as a *Homo habilis* presented difficulties. The standard view was that *Homo habilis* was an intermediate species between australopithecines and *Homo erectus*. Yet Lucy's child, although having the right age to be *Homo habilis*, possessed features more in common with Lucy, who had lived more than a million years beforehand. Indeed, the first detailed analysis of OH 62 (as Lucy's child was formally called) showed that on every measurement taken, OH 62 was even more apelike than Lucy. All this seemed to make nonsense of evolutionary principles.

There was a further conundrum. If OH 62 was a *Homo habilis*, then (according to the lineage accepted by most palaeoanthropologists) this tiny apelike creature living 1.8 million years ago had given rise to *Homo erectus*—which Turkana Boy had shown to be a strapping six-foot-tall species—all within the space of 200,000 years. The explanation given by Johanson and White was that there had indeed been a sudden change. 'Something extraordinary was happening during those few hundred thousand years of evolution, some shift in behaviour that would quite suddenly transform a *Homo* with a comparatively small brain in a primitive body into one with a big brain', Johanson wrote in *Lucy's Child*. 'An unprecedented evolutionary event had taken place'.

But others disputed this line of reasoning. Leakey and Walker believed that OH 62 was some type of gracile australopithecine, as yet unnamed. Part of the problem, they argued, was the way that palaeoanthropologists had turned *Homo habilis* into a wastebasket species, throwing in an odd assortment of fossils around 2 million years old, with a variety of morphologies and brain sizes.

Whatever classification was used, the fossil evidence was beginning to indicate to palaeoanthropologists that in the million years after 2.5 million years ago, there had been no simple linear transition from one species of *Australopithecus* to a successor species of *Homo* but rather a period of wild evolutionary experimentation.

Other scientists were coming to the same conclusion.

CHAPTER 12

A DANCE THROUGH TIME

WHILE STUDYING the fossil record of African antelopes and other bovids during the 1970s, Elisabeth Vrba, a South African palaeontologist working at the Transvaal Museum, discovered evidence of dramatic evolutionary change in their history that had occurred about 2.5 million years ago. In sudden profusion, new species had emerged; old ones had died out. The cause, Vrba believed, was a marked change in vegetation cover. Antelope species with a general diet—such as impala—survived by adapting to different vegetation; the same impala species had remained unchanged for 3 million years, thriving in a vast expanse of eastern and central Africa. But antelopes with specialised diets—such as a large variety of wildebeest, hartebeest and blesbok—either faced extinction or migrated to new niches, leading to an explosion of new species. At least twenty-nine species adapted to forest habitats became extinct in the period 2.8–2.5 million years ago, while new species adapted to open, grassy terrain emerged.

The conclusion that Vrba reached was that environmental change was more likely to promote the formation of new species in groups of dietary specialists than in those groups that were generalists. What had caused such a marked change 2.5 million years ago, she maintained, was an abrupt shift in global climate. 'Specialist species replace each other as the climate changes, like partners in a gavotte—a dance through time'.

Vrba went on to develop a radical theory about the impact of climate change on evolution. Climate change had long been regarded as an important factor. Charles Darwin acknowledged that evolution speeded up when conditions changed, but he viewed climate change as a subsidiary mechanism of natural selection that served to tighten the screws on competition. 'As climate chiefly acts in reducing food, it brings about the most severe struggle between the individuals', he wrote in *On the Origin of Species.*

Modern research shows that over the last 50 million years, the earth has experienced a progressive shift towards cooler conditions. In the early stages of the Miocene, about 20 million years ago, an immense band of forest stretched across Africa, from the Atlantic coast to the Indian Ocean, giving shelter and sustenance to a wide variety of apes, among them the ancestors of all existing apes. But by the end of the Miocene, about 5 million years ago, the forests had broken up into a mosaic of landscapes including scattered woodlands and grassland savannahs. In eastern Africa, the formation of the great highland domes of modern Kenya and Ethiopia, beginning about 10 million years ago, had added to the change, preventing moist air from the Indian Ocean from passing over the wall of mountains, throwing up a rain shadow that led the once-continuous forest to shrink and fragment.

Other changes in the global climate also had an impact. Using new techniques, palaeoclimatologists concurred with Vrba that as well as the general trend towards cooler conditions, there had been abrupt falls in temperatures. A steplike drop about 14 million years ago brought about a significant extension of the Antarctic ice sheet. A further episode occurred about 6 million years ago. About 2.8 million years ago, another sharp fall in temperatures marked the onset of continental glaciation in the northern hemisphere and the formation of the Arctic ice cap. The Pleistocene epoch—between 1.8 million and 10,000 years ago—was a time of continuing climatic instability, shifting from long cold glacial episodes to warm interglacial conditions and back again. Ice sheets and other glaciers advanced and retreated periodically from the north. At the peak of the Ice Ages, nearly one-third of the earth's surface was covered by glaciers. In Europe, northern Germany and most of England were buried under ice hundreds of feet thick; in North America, the ice sheet advanced as far south as what is now New York City.

The prevailing view until the 1960s was that the tropics had been more stable than temperate zones, providing a more consistent ecological arena for human evolution. The theory holding sway—the 'Pluvial Theory'—maintained that periods of glaciation at the earth's poles had been accompanied by increased rainfall and vegetation in the tropics. But during the 1960s, new research showed that the opposite was the case: that global cooling had led to more arid conditions around the equator; and that far from being part of a stable environment, Africa's landscape had constantly been reworked by tectonic movements, volcanic eruptions and lava flows.

Vrba's theory was that abrupt changes in global climate had led to evolutionary spurts of speciation of African mammals including hominids; indeed, she argued, their response to climate change represented the principle engine of evolutionary change. When environments were stable, there was little or no evolutionary change. 'Evolution is conservative', she wrote. 'Making a new species requires a physical event to force nature off the pedestal of equilibrium'. Environmental change was thus a forcing agent in speciation: It forced species to adapt to new conditions and caused a turnover, a change in the composition of the biota, leading to new opportunities for terrestrial life. Some species responded by migrating to new areas; others became extinct or underwent speciation. Vrba called her theory 'the turnover pulse hypothesis'.

The example that Vrba chose when launching her 'turnover pulse hypothesis' in 1985 was based on events culminating at about 2.5 million years ago. A global cooling event, beginning about 2.8 million years ago, appeared to cause a relatively abrupt reorganisation of African ecosystems after about 300,000 years. It was about 2.5 million years ago, Vrba noted, that robust australopithecines first made their appearance; that stone tools showed up in the geological record for the first time; and that the first evidence of *Homo* emerged—in the form of a skull fragment found near Kenya's Lake Baringo by her Yale colleague Andrew Hill.

Other scientists identified the same period as marking a significant change in climate and vegetation. A French palaeobotanist, Raymonde Bonnefille, used fossil pollen evidence to show that the environment at Hadar (where Lucy had been found) had changed from wet woodland conditions to arid grasslands between 3 million and 2.5 million years ago. A member of the Omo Research Expedition, Hank Wesselman, reached much the same conclusion from his study of micromammal fauna (such as rodents and other small forms). Species that were adapted to moist conditions gave way to species (such as gerbils) known to occupy arid habitats.

Vrba's 'turnover pulse hypothesis' also fitted in with the theory of punctuated equilibrium that Stephen Jay Gould and Niles Eldredge had formulated in the 1970s. But whereas Gould and Eldredge had kept to orthodox Darwinism, seeing change as coming from within the species, Vrba argued that a species' evolution was driven by the changing world around it. 'Speciation' (she said) 'does not occur unless forced by changes in the physical environment'.

An American palaeobiologist, Steven Stanley, took up the theme with dramatic flourish in a book entitled *Children of the Ice Age: How a Global Catastrophe Allowed Humans to Evolve*. Stanley argued that a biological revolution had taken place as a result of climate upheaval 2.5 million years ago. It had led not only to the eventual demise of the australopithecines but had created the conditions that enabled *Homo* to emerge with a powerful brain that elevated humankind far above the rest of the animal world.

The change had occurred at momentous pace. For 1.5 million years, the apelike australopithecines had endured in Africa without evolving significantly; their brain size was only a little larger than that of apes. In the ensuing crisis, most had no chance of avoiding extinction, but one small group managed to survive. 'During an interval of perhaps a hundred-thousand years—but possibly much less—one of its populations evolved into *Homo*'. Unlike the australopithecines, early *Homo* had the wherewithal—the brain size—to make its way in the Ice Age. The label that Stanley gave for this evolutionary event was the 'catastrophic birth' of humankind.

To underpin his argument, Stanley made use of the growing body of evidence that suggested that *Australopithecus*, far from being a fully earth-bound biped, had spent a good deal of time in trees; as well as foraging for food on the ground, it possessed considerable climbing skills, retreating to trees to avoid predators such as lions. It was a highly successful mammal, but it had been caught in an 'evolutionary straitjacket', living between two worlds, one terrestrial and one arboreal. After more than 100,000 generations, it had shown little sign of evolution.

It was climate change, said Stanley—'a flip of the climatic switch'—that had closed the door on its existence, but that had also opened the opportunity for a spurt of evolution towards *Homo*. As woodland areas shrank, australopithecines were forced to abandon the habit of tree-climbing and to make longer forays across more open terrain in search of food, constantly at risk from predators. By chance, one of its populations survived to become an earth-bound biped. 'The accidental nature of our evolutionary birth is astounding', wrote Stanley.

Few scientists were willing to support the idea that the driving force of climate change was enough to explain the emergence of humankind. In a statistical analysis examining the relationship between evolutionary and climatic patterns published in 1993, Robert Foley, a Cambridge biologist, concluded that climate was an important element but not the only one. 'Another such element is competition, both within and between species'.

> It is probably the case that where climate change is an important factor, it operates through competition. Climatic change will alter the nature, abundance and distribution of environments and resources within those environments. This will lead to changes in competitive relationships between and within species, and it is these altered competitive relationships that are likely to lead to evolutionary consequences. The consequences might be extinction of populations as the most direct effect, or speciation as a less direct one arising either out of reduced intra-community competition or the opening up of new ecological opportunities. Competition, therefore, is always likely to be the immediate cause of evolutionary change, played out within a framework determined by, among other factors, the climate.

Climate change alone, Foley concluded, was not sufficient to explain the patterns of speciation and extinction among hominids and other African primates. 'Clearly competitive relationships can change independent of climate; or alternatively, the impact of climate change will vary markedly with geographical factors. In other words, it is probable that species appear and disappear as a result of local competitive conditions rather than broad global patterns of climatic change'.

Further evidence emerged, however, reinforcing the view that key stages of hominid evolution had been directly affected by the impact of environmental change. Researchers conducting geological surveys of ancient lakes in eastern Africa discovered that, although the climate record there showed a long-term drying trend, it had been punctuated by short episodes of extreme variability, fluctuating from wet to dry conditions. Over the past 3 million years, giant lakes up to 1,000 feet deep and stretching for hundreds of square miles had been formed and had then vanished as the climate changed; the disappearances of the lakes had been followed by periods of severe drought. 'At one extreme', observed Mark Maslin, a leading palaeoclimatologist, 'the landscape would have been a true Garden of Eden, with beautiful freshwater lakes, beautiful shorelines and forests along the rivers. There would have been open spaces allowing early humans to exist easily, with water and lots of resources. But occasionally these quickly flipped into bone-dry periods, where it's 45 degrees centigrade in the middle of the day and no natural water resources'.

Three of these periods of extreme variability, said the researchers, had occurred around 2.5 million, 1.5 million and 1 million years ago, when global climate changes coincided with tectonic disruption in the Great Rift Valley. They had acted as a catalyst for evolutionary change, putting enormous pressure on hominids to adapt to the new environment. Some species were driven to the brink of extinction; others survived, acquiring higher brain power and new skills.

CHAPTER 13

NEW FRONTIERS

THE BOUNDARIES OF human origins were pushed back even further in the 1990s by a series of remarkable discoveries in the field. In Ethiopia, shortly after the authorities lifted an eight-year moratorium on the activities of foreign scientists, a new international expedition was formed to explore the Middle Awash sector of the Afar region, some fifty miles south of Hadar, where ancient sediments older than 4 million years had previously been identified.

An American geologist, Jon Kalb, had made extensive surveys of the area during the 1970s. Based in Addis Ababa, Kalb had initially played a major role in helping to establish the International Afar Research Expedition, but he soon fell out with Johanson and went on to launch his own group, the Rift Valley Research Mission, obtaining a government permit to explore the Middle Awash. According to Kalb, Johanson set out to wreck this rival group. Among the tactics he used was to spread rumours in Ethiopia and the United States that Kalb was working as an agent for the Central Intelligence Agency. In his account of his time in Ethiopia—*Adventures in the Bone Trade*—Kalb wrote of 'the cutthroat competition and backstabbing', the 'treachery and bloodletting' that plagued palaeoanthropological research there. Despite the progress he had made in the Middle Awash, Kalb's request in 1978 for funds from the U.S. National Science Foundation were rejected. That same year, he and his family were expelled from Ethiopia. In a report on these events to U.S. Vice President George Bush, a senior NSF official, Jerome Fregeau, observed in 1982: 'The exceedingly cut-throat level of competition in Eastern African anthropology is a long-standing problem. NSF cannot be blamed for it. . . .' Nevertheless, in 1987, as a result of legal action, the National Science Foundation was obliged to issue a public apology for its own part in the affair.

The new Middle Awash Research Group, starting fieldwork in 1991, was led by Tim White from the University of California at Berkeley, and an Ethiopian geologist, Giday WoldeGabriel, from the Los Alamos National Laboratory in New Mexico. Exploring an area near the small village of Aramis in 1992, a Japanese member of the team, Gen Suwa, saw the glint of a hominid molar lying amidst a mass of pebbles. In the days that followed, other bits of teeth and bone were uncovered. Among them was the lower right jaw of a child, with a milk molar still attached. It was spotted by Alemayehu Asfaw, the Ethiopian researcher who had discovered the first jawbone of Lucy's species at Hadar in 1974. Its significance was quickly realised. The milk molar was so primitive that it clearly belonged to a species that was older than Lucy's species. In all, bone fragments from seventeen individuals were retrieved, including the base of a skull and three arm bones.

Announcing their discovery in the columns of *Nature* in 1994, the Middle Awash group reported that they had found remains of the earliest known hominid, dating back 4.4 million years, nearly 1 million years older than the oldest known specimen of *afarensis*. The provisional name they gave to it was *Australopithecus ramidus*, drawn from the Afar word *ramid*, meaning 'root'. *Ramidus* was described by White as 'the most apelike hominid ancestor known'. It shared many traits with chimpanzees but it also possessed features—notably its teeth—

that had evolved after chimpanzees had split from the line leading to humans and that tied it to later hominids.

As well as describing the fossils, the Middle Awash group analysed the environment inhabited by *ramidus*. Their evidence was that it was well-wooded terrain, teeming with forest-dwelling wildlife; notably absent were savannah species. The possibility that early hominids were bipedal forest-dwellers rather than residents of the savannah was at odds with conventional wisdom. The common view was that they had adopted upright walking after emerging from the cover of trees to enable them to forage over greater distances on the savannah. But what was missing from the Middle Awash group analysis was conclusive evidence that *ramidus* in fact walked upright—the defining characteristic in modern times of the human lineage.

Back in the field at Aramis for another season, two months after the *Nature* report was published, the Middle Awash group made another remarkable *ramidus* discovery. Crawling up an embankment where recent rains had eroded the surface, an Ethiopian graduate student, Yohannes Haile-Selassie, spotted two pieces of bone from the palm of a hand. In the following weeks, the team uncovered the partial skeleton of a single individual—pieces of a pelvis; leg, ankle and foot bones; arm, wrist and hand bones; a lower jaw with teeth; and a crushed skull; in all, 125 bone fragments. Over the course of three subsequent field seasons, they found body parts of at least thirty-five other individuals.

Because of the marked differences between *ramidus* and the australopithecines, in 1995, White and his colleagues dropped the genus *Australopithecus* and substituted a new genus—*Ardipithecus*—drawing on the Afar word *ardi,* meaning 'ground'. Critics, however, remained sceptical about the claims of *Ardipithecus ramidus* to bipedal status.

It took White's team fifteen years of painstaking reconstruction and analysis before they could reach a definitive conclusion. Naming the skeleton 'Ardi', they announced in 2009 that it belonged to a female, standing four feet tall, weighing about 110 pounds and able to walk upright on flat feet in a primitive manner. Ardi's long arms and fingers and opposable big toes meant that she was also adept at climbing trees and moving through the forest canopy to reach food, to sleep in nests and to escape predators. 'In *Ardipithecus*' said White 'we have an unspecialised form that hasn't evolved very far in the direction of *Australopithecus*'.

Analysis of the teeth of *Ardipithecus* pointed to a varied, omnivorous diet of fruit, roots, insects, eggs and perhaps small mammals. Of particular significance, both male and female *Ardipithecus* had small incisors and canines, suggesting that unlike chimpanzees, baboons and gorillas the male did not bare its teeth in battles over females and was already part of a more cooperative social group. 'In all the great apes—and that includes fossil and modern—the large, tusk-like, projecting, shearing canine teeth are used as weapons, and in most of them the main use is in males fighting with other males for access to estrus females', said White. 'The earliest hominids lack that adaptation, showing much smaller canines that are not at all chimpanzee-like'. He concluded: 'Natural selection has led to the reduction of this male canine tooth very, very early in time, right at the base of our branch of the family tree'.

Another spectacular *Ardipithecus* find was made by Yohannes Haile-Selassie in 1997. Exploring the Alayla Basin on the remote western margin of the Middle Awash, he spotted a piece of jawbone lying among basalt cobbles. It came from sediments that were 5.8 million years old, making it 1.4 million years older than the Aramis skeleton. Over the next four years, Haile-Selassie and other members of the Middle Awash team collected a total of eleven hominid fossils from five individuals ranging in age from 5.8 million years to 5.2 million years, but they failed to find a skull or intact limb bones. A 5.2-million-year-old toe bone at a site near Aramis, however, provided some evidence that *Ardipithecus* at that stage might have walked upright. Six teeth uncovered in 2002 added further vital clues, indicating that although *Ardipithecus* was closely related to chimpanzee ancestors, it had started to evolve towards the human lineage. Originally considered to be a subspecies of *ramidus*, in 2004 this collection of fossils was considered to be sufficiently distinctive to be accorded its own species title: *Ardipithecus kadabba*. '*Ardipithecus kadabba*', said Haile-Selassie, 'may represent the first species on the human branch of the family tree just after the evolutionary split between lines leading to modern chimpanzees and humans'.

Alongside the Ethiopian discoveries, researchers in Kenya were also making headway. A new phase of exploration in western Turkana began in 1989, led by Meave Leakey, the head of palaeontology at Kenya's National Museums, who had taken charge of fieldwork after Richard Leakey was appointed to manage Kenya's wildlife parks. The first site she chose to explore was Lothagam Hill, a vast slab of rock heaved upwards at an angle, exposing sediments that were between 8 and 4 million years old. Once, Lothagam had stood out as an island in a much larger Lake Turkana. A combination of volcanic activity, sedimentation and time had transformed it into a palaeontologist's delight. Over the course of five years, Leakey's team collected thousands of fossil specimens: extinct hippos, horses, rhinoceroses, monkeys, giraffes, birds and a huge array of fossil fish—crabs, turtles, crocodilians and rare cichlids. The diversity of animal life was remarkable. Compared to the arid landscape of modern times, the ancient rivers and lakes of the area had teemed with life. There were, however, few signs of human remains—just six hominid teeth were found during five years of searching.

In 1994, Leakey moved her team to Kanapoi, an ancient valley south of Lothagam. During the 1960s, a Harvard professor, Bryan

Patterson, had made significant discoveries there. In 1965, he spotted a hominid elbow bone that was subsequently dated as about 4 million years old; it was the oldest known fossil at the time, predating anything that Louis and Mary Leakey had found at Olduvai. In 1967, Patterson uncovered part of an australopithecine jaw at Lothagam that was thought to be at least as old as the Kanapoi specimen. But both fossils were too fragmented to provide much conclusive evidence about early hominids.

Two weeks after setting up camp at Kanapoi, a member of the Hominid Gang, Wambua Mangao, found an upper jaw with three teeth. In further excavations, other fossils were uncovered, including parts of a shinbone that indicated upright walking. Announcing their finds in 1995, Leakey concluded that her team had discovered a new type of australopithecine, about 4 million years old, older than any previously known species. She named the new species *Australopithecus anamensis* after the Turkana word *anam*, meaning 'lake'. The size of a chimpanzee, it possessed a mixture of features combining humanlike limbs with relatively apelike jaws and teeth, more primitive than Lucy's. Leakey believed that it was the first of the australopithecines; intermediate in morphology and age between the 4.4-million-year-old *Ardipithecus ramidus* and the more advanced 3.2-million-year-old Lucy. As well as walking upright, *anamensis* retained a tree-climbing ability. The team's anatomist, Alan Walker, pointed out: 'These finds demonstrated that all parts of the human body did not evolve simultaneously but rather in bits and pieces, much like a mosaic'.

Another breakthrough occurred in 1999. Exploring sediments along the Lomekwi River, a member of the Hominid Gang, Justus Erus, found fragments of a skull that turned out to belong to a 3.5-million-year-old hominid that no one had previously encountered. It took Leakey's team a whole year to reassemble the skull from bits and pieces. Even though the skull was distorted and badly abraded, Leakey was convinced that its features set it apart from its only known contemporary, *Australopithecus afarensis*—Lucy's species. What was most striking about it was the forward position of its large, flat cheekbones relative to the jaws. Whereas Lucy's protruding face resembled that of a chimpanzee, the flat face of Leakey's specimen seemed to foretell the look of later hominids. Leakey decided therefore to assign the discovery not just to a new species but to a new genus—*Kenyanthropus platyops*—'flat-faced man of Kenya'. Quite where *Kenyanthropus platyops* fitted into the rest of the human family tree was uncertain. Meave Leakey claimed that it shared features with Richard Leakey's 1470 skull, otherwise known as *Homo habilis* but more recently named *Homo rudolfensis*, perhaps a direct ancestor. Critics, such as Tim White, argued that it was simply a variant of *afarensis*.

Nevertheless, the clear possibility remained that there were at least two different bipedal human ancestors living at the same time more than 3 million years ago. Palaeoanthropologists saw this as evidence of 'adaptive radiation', the rapid diversification of species—splitting again and again—that occurs after an initial evolutionary innovation. Although 'adaptive radiation' had long been accepted as a fundamental principle of animal evolution (as shown, for example, by the proliferation of the cat family), when it came to human evolution, many experts had preferred to stick to the notion of linear development. 'We used to look for simplicity because we wanted a neat picture', observed Daniel Lieberman, a George Washington University palaeontologist. Since the 1970s, there had been a growing body of evidence to suggest that more than one species or lineage of hominid had existed simultaneously—but only from 2 million years ago. Lucy had been portrayed as the single common ancestor. But the discovery of *Kenyanthropus platyops* pushed back the boundaries of multiple species to at least 3.5 million years. 'It shows that our past is like that of any other mammal', said Leakey. '[We have] a very complicated diverse past with lots of different species; many of which became extinct'.

No sooner had *Kenyanthropus platyops* appeared on the stage than another remarkable discovery was made by researchers exploring the Tugen Hills, a sixty-mile stretch of rugged terrain west of Lake Baringo in central Kenya. The fossils discovered there dated back 6 million years—the oldest bones yet known. But the occasion was remembered as much for an outbreak of vicious rivalry among palaeoanthropologists as for the discovery itself.

The Tugen Hills, a huge fault block standing in the middle of the Rift Valley, had long excited the interest of geologists and palaeontologists. Its layers of sediments provided a window on the past stretching back 16 million years. Since expeditions had first been launched there in the 1960s, thousands of animal fossils had been discovered.

In 1974, a thirty-one-year-old geologist, Martin Pickford, spotted a hominid molar in the 6-million-year-old Lukeino Formation, but its significance was overshadowed by the discovery of Lucy that same year. Pickford had an aptitude for rough terrain. Although born in Britain, he had spent much of his childhood on a Kenyan farm. While at school in Nairobi, he had befriended Richard Leakey. As a geology student, he worked in the Tugen Hills over a period of eight years during the 1970s, earning a doctorate from the University of London. Many colleagues, however, found his blunt manner and aloofness difficult to take. For his part, Pickford appeared to relish his role as an outsider.

In 1980, a British-born Harvard professor, David Pilbeam, was awarded a grant to establish a new research enterprise covering the Tugen Hills and the adjacent valley around Lake Baringo—the Baringo Paleontological Research Project. Pilbeam chose as field director Andrew Hill, an affable British-born geologist who had first

Mary Leakey, seen here at Laetoli at the end of a trail of hominid footprints fossilised in volcanic ash. The trail dates from 3.6 million years ago and shows that hominids had acquired the upright, bipedal, free-striding gait of modern humans by that time.

25

A single adult footprint from the trail at Laetoli. The footprints show a well developed arch to the foot and no divergence of the big toe.

Mary Leakey at work with the American palaeoanthropologist Tim White during excavations at Laetoli in 1978.

27

Donald Johanson and Tim White announce the naming of a new hominid species *Australopitchecus afarensis* in 1979, provoking a furious controversy.

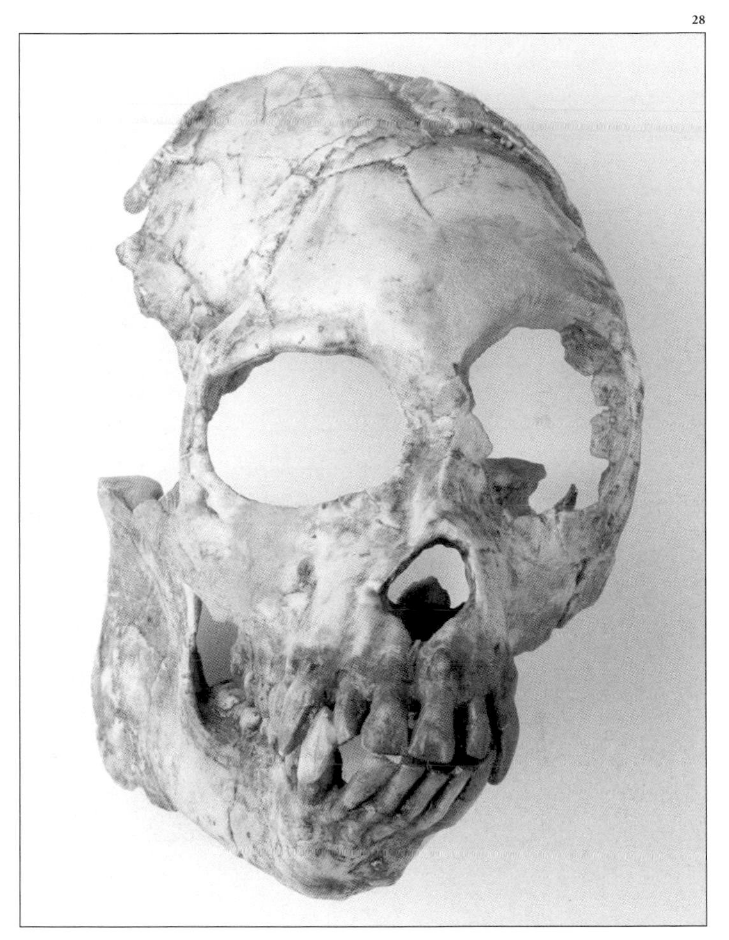

Proconsul africanus: The reconstructed skull of a Miocene-era primate discovered at Rusinga Island, Kenya.

29

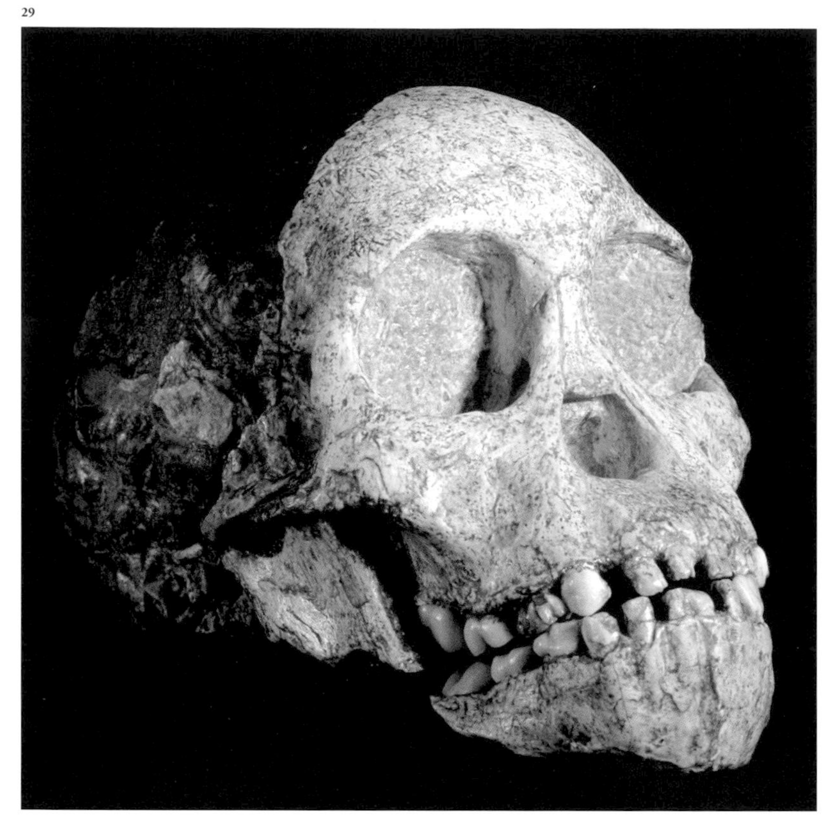

The Taung child skull—*Australopithecus africanus*—discovered in 1924.

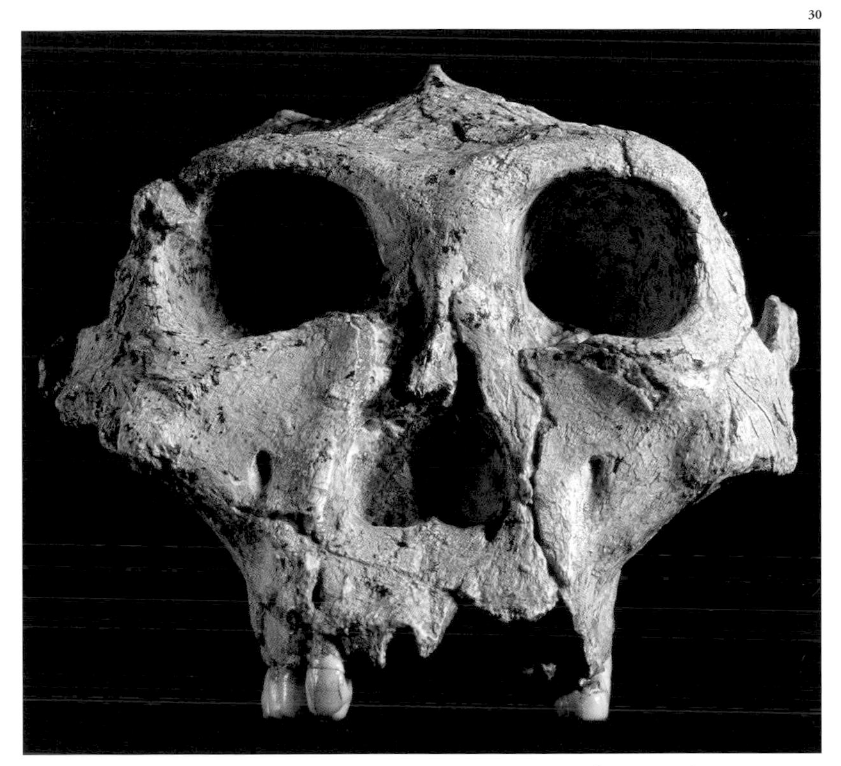

An *Australopithecus robustus* skull discovered at Swartkrans, South Africa, in 1949.

31

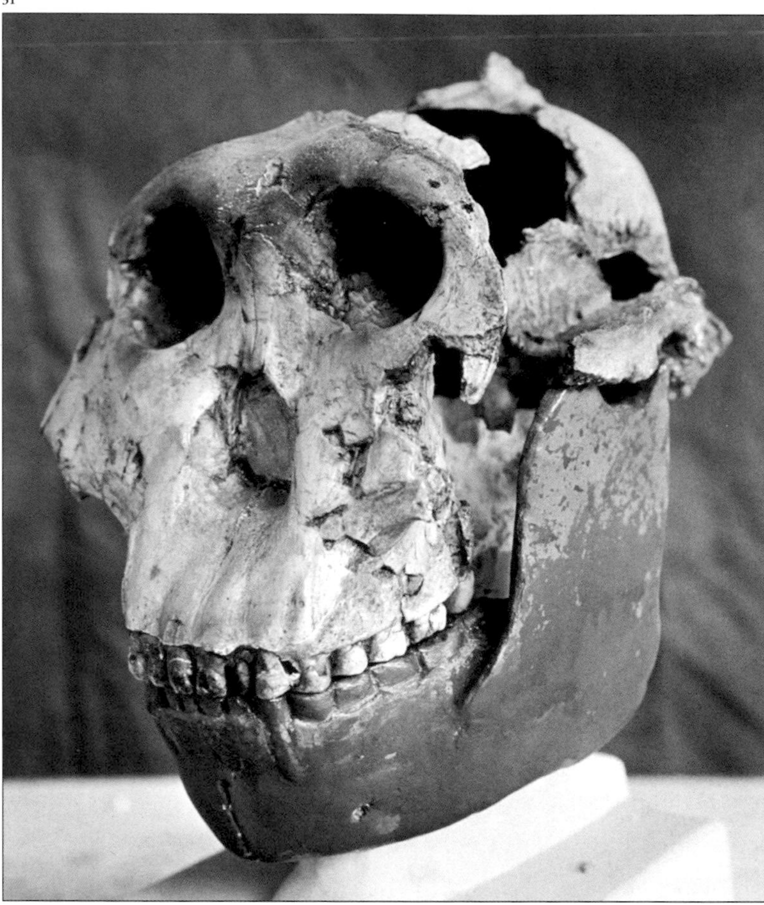

Nutcracker Man, or *Zinjanthropus boisei,* a robust australopithecine discovered at Olduvai by Mary Leakey in 1959.

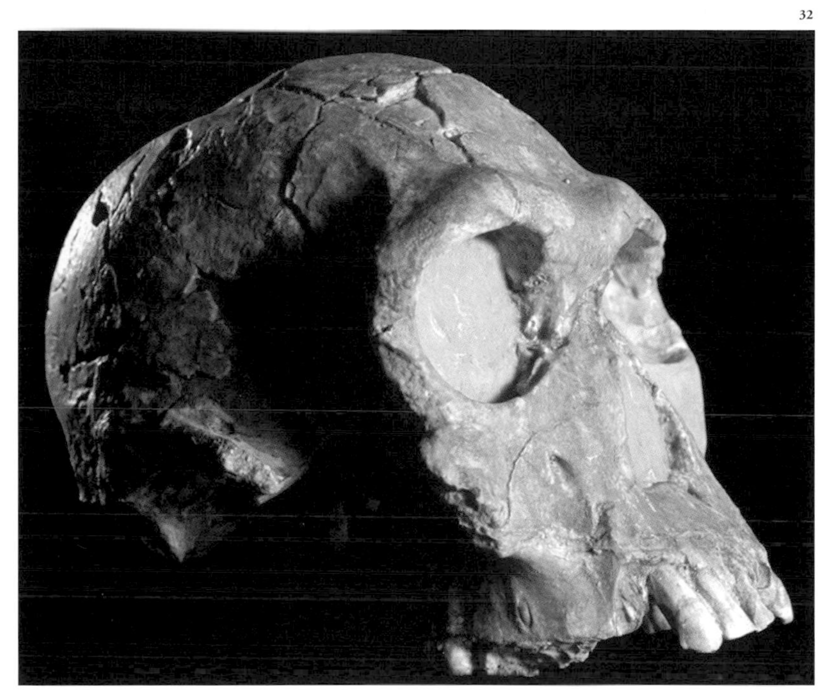

A *Homo habilis* skull found in East Turkana, Kenya, in 1973.

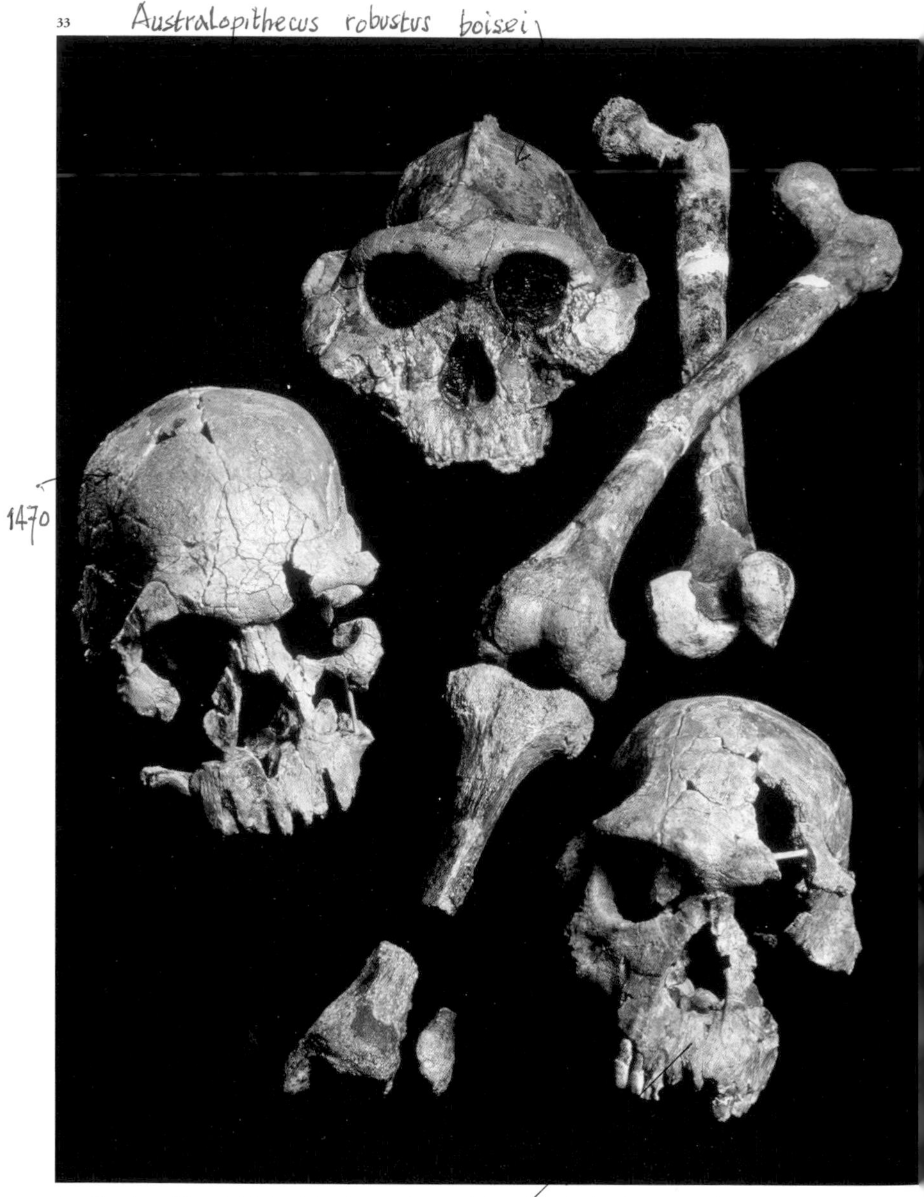

A collection of hominid fossil skulls, together with femur and tibia fragments, found at East Turkana. The skulls are: a *Homo habilis* specimen known as 1470 after its classification number (*centre*); *Australopithecus africanus* (*bottom*); and *Australopithecus robustus boisei* (*top*). These three hominid species were found among deposits that showed they lived contemporaneously, approximately 1.5 million years ago.

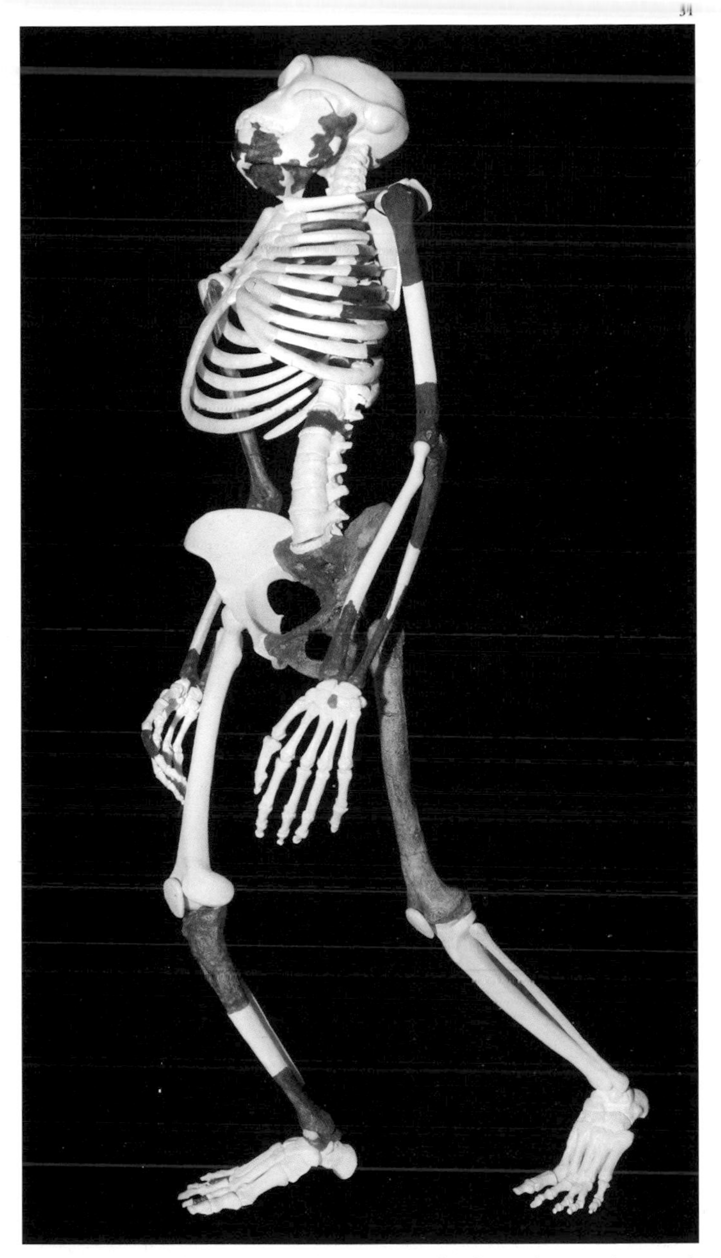

A reconstruction of the skeleton of Lucy—*Australopithecus afarensis*—discovered at Hadar, Ethiopia, in 1974.

35

A reconstruction of Lucy's head.

36

Elisabeth Vrba, a South Africa–born palaeontologist who discovered evidence of dramatic evolutionary change occurring 2.5 million years ago while studying the fossil record of African antelopes.

37

Yves Coppens, a French palaeoanthropologist involved in major discoveries in Ethiopia and Kenya, pictured in 1983.

38

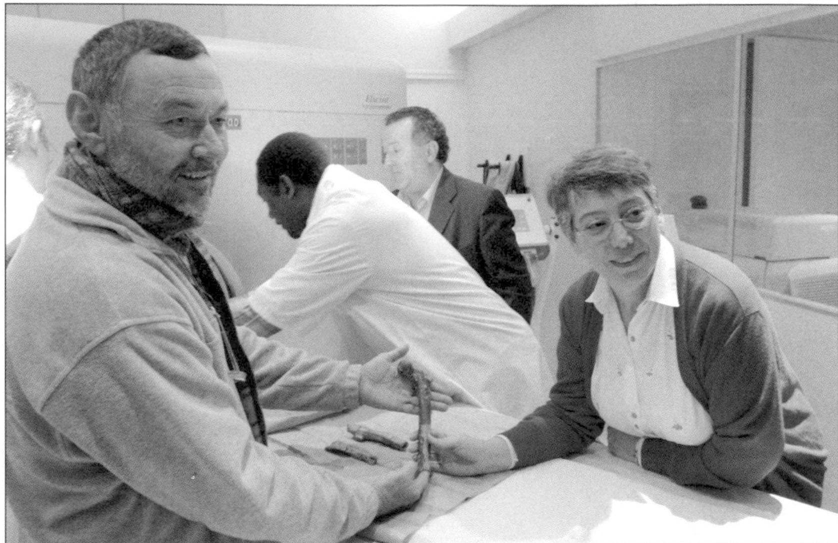

Martin Pickford and Brigitte Senut with the remains of 'Millennium Man', the first major discovery of the 21st century, which they subsequently named *Orrorin tugenensis*. p.139

39

p.147 Ron Clarke, a British-born palaeontologist, pictured alongside Little Foot, a Sterkfontein australopithecine which he discovered after a brilliant piece of detective work. pp. 148–150

40

Phillip Tobias, a South African scientist, renowned for his work at Sterkfontein and other southern African sites, pictured in 2006 at the age of eighty.

41

French palaeontologist Michel Brunet with the 7 million years-old Toumaï skull from Sahara.

An engraved ochre tablet found at Blombos Cave, South Africa, dating back about 75,000 years ago, evidence of early artistic endeavour from Africa. See p. 182

worked in the Tugen Hills as a student in 1968. The appointment was approved by Richard Leakey, then the director of Kenya's National Museums, who had ultimate control over the granting of research permits. Despite Pickford's years of research in the Tugen Hills, he was not invited to join the project. Pickford was given a post as a geologist at the National Museums, but he felt aggrieved at being excluded from the Tugen Hills project and from palaeontological work. In 1984, his contract was not renewed.

The following year, Pickford embarked on a project to explore fossil sites in Uganda previously surveyed in the 1950s by a renowned British geologist, William Bishop. He was accompanied by a French palaeontologist, Brigitte Senut, from the National Museum of Natural History in Paris. On their way to Uganda, they decided to stop off in Nairobi to study Bishop's notebooks, which were held by Kenya's National Museums.

It was a fateful visit. Leakey accused Pickford of attempting to steal the notebooks. Pickford denied the charge, but the Museums' trustees voted to ban Pickford from working on the premises, effectively ending his career prospects in Kenya. Pickford's efforts to clear his name were to no avail.

It took Pickford ten years to wreak revenge. In 1995, he launched a vitriolic attack on Leakey in a book, written in collaboration with a Kenyan colleague, entitled *Richard E. Leakey: Master of Deceit*, accusing him of every kind of malpractice. Pickford's co-author, Eustace Gitonga, was a former employee of the National Museums, the head of its exhibits department, whom Leakey had fired after accusations that he had misused funds. The book's dedication read: 'This book is for the victims of the fantastic Richard Leakey manipulations, most of whom didn't know what hit them'.

Pickford and Gitonga next plotted to outmanoeuvre Leakey and the museum establishment by setting up their own museum enterprise.

With Pickford's help, Gitonga founded the Community Museums of Kenya in 1997, persuading government politicians to grant it legal status equal to the National Museums with the right to issue its own research permits. 'No longer is paleontology in Kenya the monopoly of a single family or institution', Gitonga wrote in a letter to the journal *Science*. 'Kenyans have recuperated their heritage'. He added: 'Scientists of good faith from anywhere in the world now have a choice to carry out research in Kenya, something that was not possible for the first 35 years of the country's independence'.

Gitonga lost no time in proposing research permits for Pickford and in October 1998 the Kenya Paleontology Expedition (KPE) a joint venture between the Community Museums and Pickford's institution, the Collège de France in Paris was duly authorised to carry out research in three Kenyan provinces, including the Tugen Hills where Andrew Hill's Baringo Paleontological Research Project had been at work continuously since 1981. When Andrew Hill from his base at Yale University learned that Pickford had begun work in the Tugen Hills, he immediately protested. The National Museums supported his protest and Pickford's permit was revoked by government officials.

But not only did Pickford continue to work in the Tugen Hills, in March 2000 he invaded Meave Leakey's site at Kanapoi, seeking evidence to undermine her claims about *anamensis*. Word of his arrival there soon reached Richard Leakey at the time the head of Kenya's civil service He immediately instigated Pickford's arrest on the grounds that he had been collecting fossils illegally and intended to export them to France. On leaving Kanapoi, Pickford was intercepted by police and held in prison for five days, charged with illegal excavation. In April, however, prosecutors dropped the case. Claiming that the whole affair had been nothing more than an attempt at intimidation, Pickford and the Community Museums sued the Kenya government, the National Museums of Kenya and Leakey, alleging unlawful arrest, false imprisonment and malicious harassment. 'All of this business, all of these dirty tricks, come from Richard Leakey', Pickford told a journalist in Paris. 'It is all about egotism and power and keeping a grip on what he sees as the family business. Ever since the 1960s, the Leakeys have believed that they have exclusive rights to control all palaeontological exploration in Kenya'.

Despite the furore, members of the Kenya Paleontology Expedition soon resumed work in the Tugen Hills. In October 2000, a renowned fossil-hunter, Kiptalam Cheboi, found fragments of a jawbone with teeth attached lying on the surface in a remote area called Kapsomin. Two weeks later, Pickford and Senut arrived on the scene and recovered in all twelve more fossils from at least five individuals, including a thighbone, from four sites in the Lukeino Formation.

At a press conference in December, Pickford and Senut triumphantly announced they had found the earliest known member of the human family, calling it the 'Millennium Ancestor'. 'Not only is this find older than any other previously known, it is also in a more advanced stage of evolution', said Pickford. The thighbone showed that the hominid—the size of a modern chimpanzee—had strong back legs which enabled it to walk upright. 'Preliminary studies of the arm and finger bones reveal that the Kapsomin hominid was an agile climber in the trees, whereas its leg bones indicate that when it was on the ground it walked on two legs'.

Pickford and Senut provided more detail in two papers published in 2001. The 6-million-year-old hominid, they said, represented an entirely new genus and species. They named it *Orrorin tugenensis*, drawing on the Tugen name for 'original man', and claimed it was a direct ancestor to modern humans. Its thighbone was more human-like than that of Lucy's and other australopithecines; and its molar teeth were smaller than those of many subsequent hominids. In a simple phylogenetic diagram, they relegated Lucy and the other australopithecines to a dead-end branch, while *Ardipithecus* was moved out of the human family altogether into a line of apes. Given the age of *Orrorin*, they said, the divergence between apes and humans probably took place between 9 and 7 million years ago.

In a carefully worded comment in *Nature*, two palaeontologists from London's University College, Leslie Aiello and Mark Collard, urged caution. 'The age of *Orrorin* makes it a highly important addition to the debate about human origins. But we are a long way from a consensus on its role in human evolution'. Other palaeontologists were more critical, suggesting that the fossils belonged to a chimpanzee or one of its ancestors. To call *Orrorin* a hominid, said Bernard Wood of George Washington University, meant 'you have to rewrite human evolutionary history'.

After an extensive examination of the fossils, two American palaeoanthropologists, Brian Richmond and William Jungers, concluded that there was 'convincing evidence' to show that *Orrorin* had stood and walked on its hind limbs, providing the earliest known example of bipedal locomotion. Their analysis of its hand and arm bones indicated that it also climbed trees, 'presumably to forage, build nests and seek refuge'. But they disputed claims that *Orrorin* was a direct ancestor of the human line bypassing the australopithecines. They found instead a close similarity between the *Orrorin* thighbone and hip mechanics and those of the australopithecines.

What this meant, they said, was that the basic pattern of bipedal walking appeared very early on in human evolution and persisted with only minor variations over a period of 4 million years. The walking mechanism used by *Orrorin* had remained largely unchanged until the rise of *Homo* 2 million years ago.

Less than a year after *Orrorin*'s remains had been found, an even older fossil was uncovered in Chad in north-central Africa—more than 1,500 miles from the Rift Valley. Chad had interested French scientists since the 1960s. One of the first palaeontologists to search for fossils there was Yves Coppens. In 1960, he began exploring sandstone formations in the Djurab Desert that had turned to rock 7 million years ago; although the region is often swept by blinding sandstorms, for brief moments of time, ancient fossils there are exposed by wind stripping the surface. The following year, Coppens's wife, Françoise, discovered part of a skull they called *Tchadanthropus uxoris*, thought to be a specimen of *Homo erectus*, about 1 million years old. But civil war in Chad meant that Coppens had to turn his attention elsewhere. In 1967, he joined the Omo expedition and later the Hadar expedition that found Lucy.

As director of the Musée de l'Homme in Paris during the early 1980s, Coppens developed the theory—known as the 'East Side Story'—that the split between apes and hominids had occurred as a result of changing rainfall patterns some 8 million years ago caused by the formation of the Rift Valley, leaving apes to the west living in their natural habitat in dense equatorial forests while forcing apes to the east to adapt to more open terrain. 'Our common ancestors, the apes, were divided into a large western group and a small eastern one', he said. 'The western apes, in their forested niche, could have been ancestors of chimps and gorillas, while the eastern apes, having to adapt to the new open grasslands, could very well have been the precursors of hominids, *Australopithecus* and *Homo* in succession'.

The East Side Story became a popular theory among palaeoanthropologists until 1995, when members of a new expedition exploring the Djurab Desert—the Mission Paléoanthropologique Franco-Tchadienne (MPFT)—found the lower jaw of an australopithecine aged between 3 and 3.5 million years near Koro Toro—the first discovery of a hominid fossil west of the Rift Valley.

The expedition was led by a fifty-four-year-old French palaeontologist, Michel Brunet, who had previously hunted for fossils in the rain forests of Cameroon without success before turning to Chad. Brunet named the hominid *Australopithecus bahrelghazali*, describing it as a western cousin of Lucy. The cradle of humankind, he told a press conference in Paris, clearly covered a far wider area than had hitherto been thought.

The discovery of *bahrelghazali*—river of the gazelles—prompted further exploration of the Djurab Desert. In 2001, at a site named Toros-Ménalla, a young Chadian researcher, Ahounta Djimdoumalbaye, found a well-preserved and nearly complete cranium of an unknown hominid. Two lower-jaw fragments and three isolated teeth added to the find. From animal fossils found alongside the skull—ancient elephants and extinct pigs—it was dated at nearly 7 million years old—twice as old as Lucy. It displayed a combination of primitive and derived characteristics clearly showing it was not related to the ancestors of chimpanzees or gorillas, but living close in time to the last common ancestor between chimpanzees and humans. The skull had an elongated shape with a short, vertical face and with a massive brow ridge; the eyes were set far apart from each other; the front teeth were relatively small. The brain was estimated to be 360 cubic centimetres, below the average for all three living great apes but well within their range of variation. The positioning of the foramen magnum—the opening at the base of the skull through which the spinal cord passes—suggested that it could have walked upright on two legs, but no limb, hand or foot remains were discovered to help ascertain this.

Brunet and his colleagues concluded that the skull represented a new kind of hominid deserving a genus of its own. They called it *Sahelanthropus*—'man from the Sahel'—the name of the region bordering the southern Sahara Desert; and added the species name *tchadensis*.

Critics argued that *Sahelanthropus* was more likely to be related to chimpanzees than hominids. Martin Pickford and Brigitte Senut the discoverers of *Orrorin tugenensis*) suggested that its features were consistent with a knuckle-walking proto-gorilla.

Nevertheless, *Sahelanthropus* became a new landmark on the frontiers of human evolution. It became commonly known as *Toumaï*, a word in the local Goran language meaning 'hope of life'. It was a name given to children born in the desert before the hot, dry season began when the chances of survival diminished.

The flurry of new hominid discoveries required some rethinking about the origins of bipedalism. Ever since Darwin's time, it had been more a matter of inspired speculation than scientific evidence. Darwin averred that hominids had assumed a bipedal posture to free their hands for fashioning tools and performing other activities that in turn provided further stimulus to intelligence; the apes, meanwhile, had been trapped by the continued use of their hands as a means of locomotion among trees.

Dart, on the basis of his observations about *Australopithecus africanus*, developed the theory that upright walking had evolved in open savannah country, 'where competition was keener between swiftness and stealth, and where adroitness of thinking and movement played a preponderating role in the preservation of the species'. Essentially, Dart argued, australopithecines had to learn to live by their wits and had taken to two legs to avoid predators. Forests, meanwhile, had afforded apes with 'an easy and sluggish solution' to the problems of existence.

The 'savannah hypothesis' took root among the scientific community, offering what seemed to be an elegant explanation for the origin of bipedalism. The consensus was that it was the shift from life in the forest to life on the savannah that had set hominids apart from apes, forcing them to move upright on two legs. Once they had emerged from the cover of trees, bipedalism enabled them to scavenge or hunt game or forage for food more easily, to walk more efficiently over long distances, to peer over high grass, to carry infants, to escape predators. All the key phases of human evolution appeared to have taken place in the open grasslands of Africa.

The flaw in the savannah hypothesis became increasingly noticeable as fieldworkers provided ever more evidence that the earliest bipedal hominids occupied wooded terrain, not open grasslands. The landscape at Aramis (where *ramidus* lived 4.4 million years ago) was a mosaic of woodlands with thick underbush, flooded grasslands and swamps. The Tugen Hills site (where *Orrorin* lived 6 million years ago) was a similar mosaic of forests and woodlands with dense undergrowth and wet grasslands. The Djurab Desert site (where *Toumaï* was found) was once a lush oasis of gallery forests, swamps and wooded islets that sustained a huge diversity of animal life along the shores of a vast inland sea. The terrain there, Michel Brunet suggested, was much like today's Okavango Delta, a freshwater paradise that lies on the edge of the Kalahari Desert in Botswana.

But without the savannah hypothesis, the puzzle about the origins of bipedalism remained unanswered. One new theory that gained attention proposed that human ancestors came down out of the trees on their hind legs, already equipped to stand upright. Robin Crompton, an expert on locomotion at Liverpool University, claimed that Miocene apes, the forerunners of hominids, had evolved adaptations to their muscles and skeleton for bipedal movement while they still lived in trees. They had developed an upright stance, he suggested, as a means of foraging for fruit on outer branches. He pointed to the example of modern orang-utans which walk upright in trees when seeking outlying fruit, grabbing overhead branches to help keep their balance. This bipedal movement, he claimed, was an ancient, ancestral trait which had evolved among ape ancestors living in the forest canopy long before gorilla and chimpanzee ancestors had developed knuckle-walking for moving on the ground, and hominids had strode out into the open on two legs. 'Upright walking evolved in the ancestors of all apes, including humans', Crompton maintained. 'These techniques were later used by human ancestors to allow them to adapt to walking on two feet on the ground'.

While accepting that arboreal bipedalism was a plausible mechanism for the origins of upright walking, many scientists were sceptical about Crompton's theory. Critics cautioned against using the activities of modern orang-utans or chimpanzees as a reliable guide to the evolution of ancestors from the deep past. Orang-utan ancestors, they pointed out, were more distantly related to human ancestors than were quadrupedal gorilla and chimpanzee ancestors. They had split from the quadrupedal ape line leading to gorillas, chimpanzees and hominids some 10 million years ago, several million years before the rest of the quadrupedal line diversified.

A further intriguing piece of evidence about bipedalism came from research by Jeremy DeSilva and his team at the University of Michigan, published in 2009. After making a close study of the way modern chimpanzees in the wild scale trees—virtually vertically and with ease—and then comparing chimpanzee ankle joints with those of some thirty hominids dating back as far as 4 million years ago, they concluded that the hominids were not nearly as well adapted to climbing trees as modern chimpanzees are; they lacked the ankle structure that assists chimpanzees in climbing. Comparisons of the angle of 'dorsification'—the degree to which the ankle rotates to point the toes upwards—showed that chimpanzees are capable of a forty-five-degree bend, whereas the range in hominids was between fifteen and twenty degrees, similar to that of modern humans.

'Early hominins may have climbed trees like modern humans can and occasionally do today; however, this study suggests that vertical climbing and arboreality were not significant parts of their locomotor repertoire', DeSilva wrote in the journal *Proceedings of the National Academy of Sciences*. 'If early hominins were engaging in any substantial amount of arboreal climbing, then they were doing it in a manner . . . distinct from modern chimpanzees'.

The development of bipedalism had involved sacrificing climbing skills because the body proportion required for each was different. 'I think by 3 to 4 million years ago that trade-off was occurring', said DeSilva. 'Our ancestors were becoming very capable upright walkers, and it came at the expense of our ability to climb trees'.

Amid all the speculation about the origins of bipedalism, about the only certainty seemed to be that hominid ancestors living in tropical forests managed to develop ways of holding themselves upright in trees, passing on the advantage to generations that began to spend more time on the ground.

CHAPTER 14

LITTLE FOOT

In the post-apartheid era, South Africa joined in the run of discoveries, opening up new vistas not only on human ancestors but on modern humans. Throughout the apartheid era, while South Africa was shunned by the rest of the world, South African palaeoanthropologists and palaeontologists had persevered with their research, turning Sterkfontein and Swartkrans into the two richest cave sites in the world. By 1994, Sterkfontein alone had yielded more than 700 australopithecine specimens—males and females, infants, children, adolescents and adults—ranging in age from 3 million years to less than 1 million years. But it was a piece of brilliant detective work in the post-apartheid era that led to the most spectacular discovery of all.

In 1994, Ron Clarke, the field director at Sterkfontein, was searching through boxes of fossil animal bones collected previously from a deep underground cavern known as the Silberberg Grotto when he found several hominid foot bones among the jumble of remains that had been overlooked. The fossils had been chiselled out fourteen years before from breccia blocks originally blasted out from the lowest levels of the Silberberg Grotto by lime workers in the 1920s and 1930s.

Clarke was already renowned in the trade for his exceptional skills as a palaeontologist. Born in England in 1944, he had worked for six years as an assistant to Louis Leakey, displaying a particular talent for reconstructing hominid fossils. He had also helped Mary Leakey excavate the Laetoli footprints. He had been in charge of excavations at Sterkfontein since 1991.

Examining the bones, Clarke concluded that four of them belonged to the left foot of an *Australopithecus*. The bones revealed apelike as well as humanlike characteristics, indicating that the australopithecine had been both bipedal and prehensile, with a grasping tree-climbing capability, spending time in the trees as well as on the ground. Even on their own, the four conjoining bones represented a highly significant find; they were estimated to date back more than 3 million years, making them the oldest hominid discovery yet made in South Africa. In his initial report, Clarke speculated that the australopithecine had fallen down a narrow shaft into the underground cave system and died there, along with a variety of animals. Because the bones were so small, Phillip Tobias proposed naming the specimen 'Little Foot'.

Three years later, in May 1997, Clarke happened to open a cupboard in the hominid strong room at the Witwatersrand Medical School and noticed a box labelled 'Cercopithecoids', containing monkey fossils taken from the Silberberg Grotto. Opening the box, he immediately spotted through the plastic of a polythene bag 'a tell-tale white bone' belonging to a hominid and, upon examining it, realised that it fitted with one of the foot bones of Little Foot. Then he found three more foot bones of Little Foot in the same box.

One week later, looking through the contents of another bag, he found what proved to be a vital clue in the chain of events that followed: the slightly damaged fragment of a hominid tibia—the lower leg bone—with an oblique break just above the ankle that looked as if it could have been caused by miners' blasting. Searching for more fragments of the same tibia, Clarke retrieved—from a bag labelled bovid (antelope) tibiae—a hominid tibia shaft with the distal end intact.

This second piece of tibia turned out to fit perfectly with the other bones of the left foot of Little Foot.

Then Clarke realised that the first piece of tibia he had found came from the right leg of the same individual. He remembered finding a small foot bone in the 1994 box which, at the time, he had not been able to match with the other left foot bones. 'When I looked at it again, I found that it was a hominid bone from the right foot, a mirror image of one of the newly discovered bones of the left foot'.

In all, Clarke had accumulated twelve foot and lower leg bones from a single individual. The conclusion he reached was crucial: 'The fact that we now had foot and leg bones of both sides of one individual meant that the whole skeleton had to be there embedded in the breccia of that lower cave, the Silberberg Grotto'.

Clarke made a cast of the right tibia that had been broken off at an oblique angle and sent two of the Sterkfontein preparators (Nkwane Molefe and Stephen Motsumi) to look for a matching section of bone in the exposed breccia surfaces of the Silberberg Grotto. 'The task I set them was like looking for a needle in a haystack', recalled Clarke.

On 2 July 1997, Molefe and Motsumi began their search. The area of search included the walls, floor and ceiling. The surfaces were damp and covered in mud. The only light came from their hand-held torches. Yet on the second day they found an exact match for the cast in the side of a long slope of hard breccia. 'The fit was perfect despite the bone having been blasted apart by lime workers sixty-five or more years previously', said Clarke. 'I knew then that, encased in that steep slope of ancient cave infill, we would uncover something that palaeoanthropologists had wanted for so long—a complete skeleton of *Australopithecus*'.

Using hand-held lamps, Clarke, Molefe and Motsumi began to chisel away carefully at the concrete-like breccia. By May 1998, they had uncovered the lower legs. But then, to their consternation, for month after month, they found nothing more. Clarke eventually deduced that the upper part of the body had collapsed into a lower cavity and had been sealed over with thick stalagmite. He selected an area of stalagmite for Motsumi to work on. The next day, Motsumi made a direct hit on pieces of bone that turned out to be the back of a lower jaw and the end of an upper arm bone. Clarke cautiously chipped away more of the rock. 'The glint of tooth enamel sent a shiver down my spine. It was an upper molar which, together with the lower jaw, told us that we had located the skull'.

Further chiselling revealed that the australopithecine lay on its back on a slope with its left arm extended above its head, its right arm by its side and its legs crossed. The skull was complete, with only some minor cracking and displacement. The skeleton suggested it was an adult, about four feet tall, aged by palaeomagnetic dating as 3.3 million years old. Of particular significance was the discovery of a perfectly preserved hand. It was shaped like that of a modern human hand in proportion, but much more muscular, with a short palm, a long thumb and relatively short fingers. 'Such a hand with a long and very muscular opposable thumb had apparently evolved in our ancestors for the purpose of firmly grasping branches during tree-climbing', Clarke explained. 'This, together with the almost equal arm and leg lengths and the slightly opposable big toe, is consistent with a human ancestor that climbed cautiously in trees'.

He contrasted the australopithecine's arms and hands with those of modern apes. Modern apes possess long arms and long hands, with long palms, long fingers and short thumbs, developed so that they can suspend themselves beneath branches and move by arm-swinging from branch to branch. When walking on the ground, they use their long arms as supports and walk on their knuckles. By comparison, human ancestors like australopithecines never went through a knuckle-walking stage, but instead retained a primitive hand that was specialised only in its opposable thumb for branch-grasping.

'It was this long opposable thumb and relatively short palm and fingers that provide the necessary manual ability for tool-use and tool-making', said Clarke. 'It is the combination of this hand and the large complex brain that has enabled humans to develop into the only advanced cultural animal on earth'.

The cave systems in the Sterkfontein area continued to yield spectacular finds. In 2008, a research team led by Lee Berger, a palaeontologist at the University of the Witwatersrand, discovered the first of four fossilised skeletons in a pitlike excavation at Malapa, once part of an ancient underground cave system about 150 feet deep. The Malapa fossils—two adult females, an adolescent boy and an infant—were found within a few feet of each other, suggesting they had died at the same time or soon after one another. The theory was that they had probably died after falling or climbing down a vertical shaft leading to the underground cave system. Within days of their death, a sudden rainstorm and mudflow had washed their bodies—along with the bodies of sabre-toothed cats, hyenas, wild dogs and horses—into a deeper, waterlogged recess where they were encased in calcified rock.

The fossils were dated about 1.9 million years old. Living on the cusp of the emergence of *Homo*, they possessed a mixture of features: Some were modern, with small teeth, long legs, projecting noses, advanced pelvises; others were archaic, with long arms and small brains. Berger's team decided they should be assigned as australopithecines rather than to *Homo* and gave them the name *Australopithecus sediba*—a Sesotho word meaning 'fountain' or 'wellspring'.

Announcing the finds in 2010, Berger followed the long tradition among palaeontologists of making grand pronouncements about their field discoveries. 'We do feel that possibly *sediba* might be the Rosetta Stone for defining for the first time just what the genus *Homo* is', he said.

PART TWO

CHAPTER 15

PROTOTYPES

TEN MILLION years ago, apes were the lords of creation. Originating in Africa, more than fifty different species roamed the world during the Miocene age. Miocene apes flourished in particular in the tropical forests of eastern Africa. Among them were *Proconsul* and various contemporaries, thought to include the last common ancestor of both modern apes and hominids. Fossils found in the volcanic highlands of Kenya and Uganda reveal them as tailless, fruit-eating quadrupeds, about the size of a female baboon, which lived in trees and moved on the forest floor on all four legs.

This thriving ape community then mostly died out, probably as a consequence of global climate change affecting their environment. Climate records tell of a dramatic cooling around the world between 6.5 million and 5 million years ago. Polar ice caps expanded; sea levels plunged so low that the Mediterranean was repeatedly drained; icy winds blew off cold oceans. Africa escaped the icy conditions, but in the dry, cold climate, the Miocene forests shrank, fragmenting in places into more open woodland. Whole ape populations perished. Only a few survivors emerged: the ancestors of orang-utans and gibbons in Asia, the ancestors of gorillas and chimpanzees in Africa—and hominids.

The fossil record of this period is sparse, almost non-existent. According to molecular-clock calculations, hominids and the chimpanzee line split from a common ancestor between 8 and 6 million years ago. But the identity of this last common ancestor, its exact nature and the dates at which it lived remain obscure.

Recent discoveries have produced a number of contenders said to have crossed the hominid threshold: *Sahelanthropus* from Chad, dated between 7 and 6 million years old; *Orrorin* from Kenya, dated at 6 million years old; and *Ardipithecus kadabba* from Ethiopia, dated between 5.8 and 5.2 million years old. None of them provides conclusive evidence of habitual upright walking—the principal characteristic used to define the human lineage.

Sahelanthropus consists of a skull—an intact cranium, some lower jaw fragments and several teeth. It is the only exhibit of a skull recovered from a 5-million-year-long period. It shares several facial features with later hominids—a short face with a massive brow ridge and a mouth and jaw that protrude less than in most apes. The positioning of its foramen magnum suggests a posture similar to bipeds, with the skull balanced atop a vertically held spine. But other aspects of the skull are decidedly more apelike than other hominids. Moreover, no foot or leg bones have been found to help ascertain on which side of the divide between apes and hominids it belongs.

Orrorin consists of a collection of skeleton bones from four sites in the Tugen Hills—parts of a couple of thigh bones, part of an upper-arm bone, two jaw fragments and teeth, but no cranium. Its claim to hominid status rests mainly on its leg bones, which show features associated with upright walking. But too few fossils of *Orrorin* have been found to make any classification certain.

The claim of *Ardipithecus kadabba* to hominid status is based largely on the size and shape of its teeth, which are said to be similar to those of early australopithecines. The anatomy of an inch-long toe bone provides some indication that it was able to move its feet like a hominid. But, as with *Orrorin*, too few fragments have been found to advance its claim much further.

The evidence suggests that from an early stage, the transition from ape ancestors to hominids involved a bout of evolutionary experimentation played out over the course of several million years. Hominids emerged not just as a single species but as a collection of similar species displaying a mixture of primitive and derived features. As rain forests broke up into more open woodlands, they were forced to explore and adapt to new forest-edge and woodland habitats. Whereas ape ancestors had made occasional use of standing upright to obtain food or walking bipedally—as modern apes such as gorillas and chimpanzees do—hominids began to rely increasingly on bipedalism as a principal method of locomotion on the ground, moving over greater distances in search of food, while retaining their tree-climbing habits for safety. Bipedalism became the key physical adaptation that set the hominid line in motion.

Modern experiments have shown that humans walking on two legs use only one-fourth of the energy of chimpanzees knuckle-walking on four legs. As well as saving energy, bipedalism had the advantage of freeing hands for purposes other than supporting body weight, such as carrying food or objects. Standing upright also enabled hominids to see further over distance to spot predators. Furthermore, it may have helped reduce exposure to heat from the rays of the tropical sun in a more open environment.

The first definite evidence of upright walking comes from *Ardipithecus ramidus*, a hominid species that lived 4.4 million years ago, first discovered by Tim White's team in 1992 in the Afar region of northeast Ethiopia. A partial skeleton of a female named 'Ardi' and other fossils reveal that *ramidus* possessed a primitive walking ability using flat feet while retaining certain anatomical features such as long arms, large hands and opposable big toes that allowed it to continue a tree-living existence at the same time.

Next comes the first of the australopithecines, hominids that were better adapted to walking on the ground. *Australopithecus anamensis* is a collection of fossils ranging in age from 4.2 to 3.8 million years ago, found by Meave Leakey's team at two sites in northern Kenya, Allia Bay to the east of Lake Turkana, and Kanapoi, to the southwest of the lake. The sample of fossils is small, but it includes pieces of tibia—the lower-leg bone—that give a firm indication of upright posture. *Anamensis* fossils have also been found in Middle Awash deposits in Ethiopia.

Anamensis is now regarded as the likely progenitor of *Australopithecus afarensis*, a species dating from 3.9 million to 3.0 million years ago, found mainly at Awash Valley sites in Ethiopia. Its most famous member is Lucy, a tiny individual standing at little more than three feet tall which lived 3.2 million years ago. Another notable example is the partial skeleton of a 3.6-million-year-old male, discovered in the Woranso-Mille area of Ethiopia's Afar region in 2005, which stood at about five feet tall. *Afarensis* possesses a mixture of apelike and humanlike features. Its face, teeth and small braincase are similar to those of ape ancestors. But the shape of its pelvis and its knee joint clearly indicate that it walked erect. Locomotion experiments show that although it had not yet achieved the full potential of bipedal gait, it was nevertheless able to travel some distance upright. While it continued to use trees to forage for fruits and leaves and to seek safety from predators, it spent much of its time on the ground. Kenya's flat-faced *platyops*, dated at 3.5 million years ago, shares many features with *afarensis*.

A variety of other australopithecines then appear on the scene, including several species from southern Africa. Cave sites in Sterkfontein Valley, to the west of Johannesburg, have produced a plethora of specimens. Among the oldest is Little Foot, the Sterkfontein australopithecine found by Ron Clarke, dating back 3.3 million years, regarded as a possible antecedent to *Australopithecus africanus. Africanus* itself prospered for a period of a million years, from 3 to 2 million years ago. First named by Raymond Dart upon discovering the Taung child which lived about 2.7 million years ago, *africanus* shares more features with the later *Homo* species than any previous australopithecine. Although its brain size was similar to that of apes—about 400 cubic centimetres—its braincase was shaped like that of a *Homo*. It also had a relatively flat face and short jaw.

Another branch of the australopithecines—the 'robust' variety—crops up at sites in both southern and eastern Africa. The first evidence of their existence was discovered in 1938 by Robert Broom, who coined the name '*Paranthropus robustus*' to account for their massive chewing teeth. A later generation of palaeontologists took up the name when sorting out classification. Louis Leakey's *Zinjanthropus* became known as *Paranthropus boisei* along with other *Paranthropus* discoveries in eastern Africa. The size of their jaws and teeth is attributed to changes in diet brought on by the effects of a drier, cooler climate. They became specialised plant-eaters, adapted to eating hard nuts and seeds and large quantities of coarse, fibrous roughage low in nutritional value. Although their brain size increased to about 500 cubic centimetres, their fate was to head into an evolutionary cul-de-sac. In eastern Africa, *Paranthropus boisei* survived for nearly 1 million years, from 2.3 to 1.4 million years ago, before becoming extinct. In southern Africa, *Paranthropus robustus* is estimated to have lived between 2 and 1.5 million years ago.

As a group, the australopithecines were remarkably successful. They persevered on earth for a period of 3 million years, exploring and adapting to new habitats, developing their bipedal gait and use of hands, and expanding the range of their food sources. It was from their ranks that the first species of *Homo* emerged. Despite their small brains, they may also have been responsible for producing the first primitive stone tools—the world's first technology.

The oldest recognised stone tools—consisting of sharp flakes chipped off small cobbles and hand-sized stone hammers used to hit the cores—date back to 2.6 million years ago. They come from a site on the Gona River in Ethiopia's Awash Valley, five miles to the west of Hadar (where Lucy was found in 1974). What is notable about them is the degree of skill involved in their making. The toolmakers were selective about what types of stone they chose, collecting particular kinds of cobbles they knew would flake more easily, sometimes from distant locations. Cut marks found on animal bones at the site indicate that sharp-edged flakes were used for cutting meat from carcasses and for cracking open bones for edible marrow.

Further evidence of skilful tool-making comes from Lokalalei, a site to the west of Lake Turkana in Kenya dating back to 2.3 million years ago. Researchers there have found what they termed 'a tool factory' where toolmakers turned out hundreds of flakes, striking cobbles at the right angle and with the right force to give them effective results. The toolmakers appeared to prefer phonolite as their raw material, producing as many as thirty flakes from a single core; other types of rock were found with only a few fragments missing, suggesting they had been tested and then discarded. Archaeologists examined more than 2,000 stone fragments at the site. By piecing together some 285 flakes, they were able to reconstruct thirty-five of the original stones from which the tools had been made. The evidence indicates that small-brained hominids living about 2.5 million years ago were capable of mass production that required forward planning and learning.

The makers of these stone tools remain unidentified. No hominid fossils have been found at the sites, though parts of a 2.5 million year-old australopithecine—*Australopithecus garhi*—have been recovered near Gona. But in view of the skill the toolmakers at Gona applied, the origin of toolmaking clearly lies further back within the australopithecine world. Cut marks on bones found at Dikika in Ethiopia provide evidence of possible tool use as far back as 3.4 million years ago. Although simple, this pioneer technology was effective enough to last for nearly 1 million years, being used long after the first *Homo* had arrived on the scene.

Toolmaking opened up a vast new range of possibilities. It enabled hominids to dramatically expand their sources of food. Hitherto, they had depended on fruits, roots, herbs and insects. Now they could use stone flakes to cut up animal carcasses so they could carry the meat to safe locations. Meat and marrow fat provided a nutritious and concentrated food, nourishing brain development. And brain development was the stimulus propelling the evolution of *Homo*.

CHAPTER 16

Pioneers

The threshold between australopithecines and the first species of *Homo* is little more than a blur. The fossil record between 2.5 and 2.0 million years ago is so sparse that palaeoanthropologists have yet to determine which species of *Homo* came first and when it emerged from the ranks of australopithecines. What appears certain is that *Homo* arrived at a time when the world was experiencing another dramatic change in climate, leading in Africa to a cooler, drier environment in which forests there shrank further and savannah grasslands expanded. Added to the impact of global climate change in eastern Africa was a local bout of extreme climate variability—wild swings from wet to dry conditions followed by droughts—which put enormous pressure on hominids to adapt.

Two species have been named as contenders for the earliest examples of humankind: *Homo habilis* and *Homo rudolfensis*. But strong doubts remain about the credentials of both of them. *Homo habilis*—handy man—is the name given in 1964 to a collection of fossils found by the Leakeys at Olduvai Gorge to describe what was said at the time to be the world's oldest toolmaker, dating back 1.8 million years. Although the brain size of *habilis* (600–700 cubic centimetres) was below the level previously agreed to constitute qualification for a listing as human (at least 750 cubic centimetres), the Leakeys argued that because of its dexterity in making tools, it merited inclusion. Similar fossils subsequently recovered from other sites in eastern and southern Africa, ranging in age from 2.3 to 1.6 million years ago, have been placed in the same category.

One difficulty with *Homo habilis* is that it differs little from contemporary australopithecines: It is similar in size to them—about half the height of a modern human; and its brain capacity is generally only slightly larger. Moreover, it includes a disparate variety of fossils: Some faces are small and projecting, and others are large and flat; lower jaws also vary in size and shape.

Because of the variation, many palaeoanthropologists prefer to place the larger specimens of *habilis* in a separate category: *Homo rudolfensis*. The key exhibit is the 1470 skull found by Bernard Ngeneo at Koobi Fora in 1972 and named *Homo rudolfensis* in 1986 to commemorate its location—Lake Rudolf, the previous name for Lake Turkana. Dated as 1.9 million years old, it has a long and broad face with eyes set well apart and a single continuous brow ridge. Its braincase is relatively large, with a volume of 750 cubic centimetres. The drawback about *Homo rudolfensis* is that there are too few specimens upon which to base a judgement.

The first species to be recognised as being distinctly human appeared in Africa just under 2 million years ago. It was classified at first as *Homo erectus* because it was similar to fossils identified previously in Asia by that name, but it later became known as *Homo ergaster* (meaning 'workman' to distinguish its African origin). Palaeontologists agreed in the 1990s that there was sufficient difference between *erectus* and *ergaster* specimens to accord *ergaster* its own species designation; *ergaster* had a rounder head and thinner cranial vault bones, features that were passed down the line towards *Homo sapiens*. Some palaeoanthropologists, however, continued to refer to the species as 'early African *Homo erectus*'.

Homo ergaster occurs in the fossil record between 1.9 and 1.5 million years ago. Its prize example is Turkana Boy, the well-preserved skeleton of an adolescent found by Kamoya Kimeu in northern Kenya in 1984, which is dated as about 1.6 million years old. Turkana Boy had the build and body proportions of a modern human: He was tall, long-legged and slender, with narrow hips and shoulders, well adapted for long-distance walking on the savannah; his forearms were short and his fingers were no longer hooked. He had far less body hair than his hominid ancestors. His braincase, however, was small. If he had reached adulthood, his brain size would have been about 880 cubic centimetres, twice the size of an australopithecine brain, but only about two-thirds the size of a modern human brain.

The achievements of *Homo ergaster* were remarkable. Taking advantage of their new striding gait, groups of *ergaster* slowly expanded from their home base in eastern Africa into other areas of Africa and then beyond. A collection of hominid fossils and Oldowan stone tools found in Dmanisi in Georgia in the 1990s indicate that by 1.8 million years ago, they had reached the Caucasus. Other groups—the forerunners of *Homo erectus*—made their way to East Asia. Stone tools and fossils from China and Indonesia suggest they arrived there about 1.7 million years ago.

The first immigrants to reach western Europe from Africa travelled there via the Levant, the lands of the eastern Mediterranean. The earliest evidence of their existence comes from two sites in the mountains of northern Spain. In the mid-1990s, a team of Spanish archaeologists exploring the Gran Dolina site in the Sierra Atapuerca region, east of Burgos, uncovered the bones of six individuals dated at around 780,000 years old. Some features appear to link the individuals to *Homo ergaster*, but facial bones—in the shape of the nose and cheekbones—look more modern. The Spanish team therefore proposed an entirely new species of human that they called *Homo antecessor*, or

'Pioneer Man'. In 2007, the same team exploring a nearby cave came across another hominid fossil—a piece of jawbone with a few teeth attached—that was dated as 1.2 million years old. The first evidence of human settlement in Britain—flint tools found in Norfolk—dates back to the same period—between 950,000 and 800,000 years ago. But the attempt at colonisation ultimately failed, probably as a result of one of the severe glacial episodes that gripped Europe between 800,000 and 600,000 years ago.

Not only were early Africans adept at exploring new terrain, but they eventually devised more advanced techniques for manufacturing tools. Initially, for a period of several hundred thousand years, they relied on the same Oldowan technology that their ancestors had used as far back as 2.6 million years ago, turning out simple stone-flake tools for cutting purposes. But before fading from the African scene, *Homo ergaster* developed distinctive pear-shaped hand-axes, worked symmetrically on both sides, and pointed at one end, rounded at the other. The first evidence in Africa of this new technology was discovered during an expedition that Louis Leakey mounted in Kenya in 1928. Hand-axes had previously been identified by archaeologists working at a site at Saint Acheul near Amiens in northern France in the nineteenth century and had been named by them 'Acheulean'. But the origin of the technology lay in eastern Africa. The earliest examples of these bifacial tools, dating back to 1.6 million years ago, come from the West Turkana region of Kenya.

Hand-axes proved to be a highly successful innovation. They were portable, quick to deploy and adapted to a range of tasks from delicate cutting to hacking through tree branches. The same technology remained in use for more than 1 million years, spreading across Africa and much of western Asia as well as Europe.

Another major achievement of Africa's *ergaster/erectus* was to find a way of controlling the use of fire. In 1984, a South African archaeologist, Bob Brain, uncovered 270 fragments of charred animal bone in a cave at Swartkrans, west of Johannesburg, that showed signs of having been heated in a campfire. The bones were found in several distinct layers of limestone, dating between 1.5 and 1.0 million years ago—the earliest direct evidence of the controlled use of fire. Brain concluded that the Swartkrans fire makers had not yet discovered how to start their own fires, but had found ways of keeping alive bushfires commonly started by lightning strikes. Several sites in Kenya have also yielded clues about the early use of fire, dating back to 1.5 million years ago, but the evidence there—in the form of lumps of charred clay—is less conclusive. Other evidence of the controlled use of fire comes from a site at Gesher Benot Ya'aqov in northern Israel, where researchers have found clusters of burnt artefacts—mainly flint implements—dated to about 790,000 years ago at what was probably a *Homo erectus* camp on the shores of an ancient lake.

The 'capture' of fire was a crucial innovation. It enabled early humans to expand their diet by cooking seeds, plant foods and meat. It also provided a source of heat and light and protection from predators. 'The acquisition of fire' claims anthropologist Frances Burton, in her study of fire 'was the engine that propelled the incredibly fast evolution of humans'.

A Harvard anthropologist, Richard Wrangham, has taken the idea further. He developed the theory that it was the ability of early humans to use fire to produce cooked food that transformed human evolution. Wrangham identifies two crucial steps in the transition from ape to human. The first occurred about 2.5 million years ago when *Homo habilis* took up meat-eating, thus enhancing brain development. But the second step according to Wrangham produced far more dramatic change. It occurred from about 1.9 million years ago when *Homo ergaster/erectus* emerged on the scene. Wrangham speculates that the reason *Homo ergaster/erectus* acquired a larger body, a larger brain, smaller teeth, and a smaller stomach and guts was not just because of the meat-eating habit but principally because they had discovered how to cook food. 'Meat eating has been an important factor in human evolution and nutrition, but it has had less impact on our bodies than cooked food', he writes in *Catching Fire,* his book explaining his 'cooking hypothesis'. Cooking provided early humans with a more nutritious range of food, yielding more energy; less time was needed for digestion; more resources were freed to fuel brain development. It was cooked food that made our brains uniquely large. 'We humans,' proclaims Wrangham, 'are the cooking apes, the creature of the flame'.

Africa's *Homo erectus* is also notable for adjusting its anatomy to cope with the growing size of its brain. Whereas the average cranial capacity of *Australopithecus afarensis*—Lucy's tribe—had been an estimated 410 cubic centimetres, by the time *erectus* emerged, its brain had doubled in size to about 900 cubic centimetres, boosted by improved diets that included meat. Giving birth to bigger-brained babies required changes in female anatomy. The shape of Lucy's birth canal was too narrow to allow for the birth of large-brained babies. For many years, because of a lack of fossil evidence, scientists assumed that the pelvis of a female *erectus* had remained relatively narrow, restricting *erectus* brains at birth to be no larger than 230 cubic centimetres. But in 2001, researchers at Gona in Ethiopia—an area adjacent to Hadar, where Lucy was found—discovered the almost complete pelvis of a 1.2-million-year-old *erectus* female that produced a dramatic new perspective. Piecing the pelvis fragments together, they were struck by its unusual width. *Homo erectus,* it turned out, had performed an evolutionary leap making it capable of producing babies with brains at birth much closer in size to modern humans. The short-statured Gona female had hips that were proportionally wider than those of modern humans, with a birth canal more than 30 per cent larger than previous estimates, allowing for the delivery of a baby with a brain as large as

315 cubic centimetres; by comparison, a modern human baby's brain at birth is about 380 cubic centimetres.

Being born with a proportionally larger brain meant that *erectus* babies probably became independent far more quickly than modern human infants—a useful survival adaptation in the African savannah. The female pelvis, meanwhile, still remained narrow enough to make bipedal locomotion reasonably efficient.

By 1 million years ago, the only surviving hominid species on earth was *Homo erectus*, strung out over three continents: Africa, Asia and Europe. Mary Leakey, noting how *erectus* in Africa had kept turning out the same Acheulean hand-axes for hundreds of thousand of years, attempting no innovation, referred to *erectus* as a 'dim-witted fellow'. Yet *erectus* enjoyed a lifespan on earth longer than any other hominid species, surviving from almost 2 million years ago through periods of extreme climate variation to about 200,000 years ago and possibly beyond; one recent discovery of fossil skulls in Java placed *erectus* living there only 50,000 years ago.

An even more intriguing discovery was made on the remote Indonesian island of Flores in 2003. A team of palaeontologists excavating a cave at Liang Bua uncovered the skull and partial skeleton of a tiny human female, no more than three feet tall, with a brain size of only 417 cubic centimetres. Because of its small size, they nicknamed it 'the hobbit'. Their initial assessment, published in *Nature* in 2004, was that it belonged to a diminutive new species descended from an ancestral population of *Homo erectus* which had become isolated on Flores and had evolved into miniature form as a result of 'island dwarfism'. They subsequently found six other 'hobbits', all of them of similar size. Dating the specimens indicated that the hobbits had survived until as recently as 17,000 years ago. Sceptics argued that they were simply modern people who had suffered from some kind of pathological condition such as microcephaly, or a similar disorder

But further investigations presented more of a puzzle. Several features of the feet suggested that they were more akin to an even earlier species than *Homo erectus*—that is, to *Homo habilis*. Scientists who reviewed hobbit research at a symposium at Stony Brook in 2009 concluded that they merited inclusion as a new human species: *Homo floresiensis*. Stone tools found on Flores indicate that the island may have been occupied as far back as 1 million years ago.

About 700,000 years ago, during a period of severe climate fluctuations, a new contender appeared on the scene in Africa with an even bigger brain. Whereas brain size between 1.8 million and 700,000 years had remained remarkably stable at about 65 per cent of the modern average, it now increased to about 90 per cent. Like *erectus*, the new arrival spread rapidly to Europe. The first clue to its existence came from a gravel pit at Mauer, near Heidelberg in Germany; in 1907, workmen found a strange-looking lower jaw with a complete set of teeth which were clearly human, though the thickness of the jawbone seemed more apelike. As it was unlike any previous discovery, it was assigned a new name: *Homo heidelbergensis*. Fourteen years later, in 1921, miners cutting through an extensive limestone cave system at Broken Hill in Northern Rhodesia (now Kabwe in Zambia) came across a nearly complete skull with a massive bony brow ridge and a braincase, elongated from front to back, with a low forehead, that seemed close in size to that of a modern human. Brought to Britain, it was assigned to another new species: *Homo rhodesiensis*.

Subsequent discoveries in Africa and Europe began to point to a large-brained species intermediate between *Homo erectus* and *Homo sapiens* that survived until 100,000 years ago. In Africa, specimens have been found over a vast area ranging from the sand dunes of Saldanha near Cape Town, to Lake Ndutu in Tanzania, to Ethiopia. In 1976, members of Jon Kalb's team exploring the Middle Awash region found a partial skull at Bodo with a brain volume of about 1,250 cubic centimetres, dated to about 600,000 years ago. Scientists use both names—*heidelbergensis* or *rhodesiensis*—to describe the same species.

At the beginning of Africa's Middle Stone Age, about 280,000 years ago, *Homo rhodesiensis/heidelbergensis* achieved a further breakthrough in toolmaking, supplanting the old Acheulean technology that had endured in Africa for a million years. Hitherto, toolmakers had relied principally on chance to strike a suitable flake. Now, using a new method known now as the Levallois technique named after the suburb of Paris where examples were first found *rhodesiensis* began to produce flakes from carefully-prepared cores that were designed for specific purposes—for cutting, scraping or piercing. Another type of stone tool developed by Middle Stone Age Africans was the stone point, flakes that were retouched along both faces to make a point suitable for a spear tip.

They also made copious use of mineral pigments such as red ochre. Excavating a site near Kenya's Lake Baringo dating back 285,000 years ago, Sally McBrearty of the University of Connecticut uncovered huge quantities of red ochre, together with grindstones used to process it into powder. McBrearty believes that the inhabitants of Baringo used ochre for symbolic purposes, possibly to decorate their bodies. A site at Twin Rivers in Zambia, dating from the same period, yielded pieces of hematite and limonite—the sources of red and yellow ochre—that indicated that they had been used as chunky crayons.

In Europe, *heidelbergensis* immigrants went their own way, gradually diverging from African populations. Their range was extensive. Remains have been found in England, France, Germany, Greece, Hungary and Italy. A tibia (shinbone) and two teeth dug up at Boxgrove in southern England date back to around 500,000 years and represent the oldest human known from the British Isles. The tibia suggests that it belonged to a heavily built, muscular individual nearly six feet tall. Skulls and skeletal bones from the Spanish site at Atapuerca paint a similar picture. Some individuals there had brain sizes falling in the modern human range of 1,200 to 1,500 cubic centimetres. Wooden spears dating back 400,000 years ago, found buried in a peat bog in Schöningen in Germany, show that *heidelbergensis* groups were involved in hunting large animals. They had also begun to construct rudimentary shelters. Sites at Terra Amata in southern France and Bilzingsleben in Germany have yielded evidence of foundations of large oval huts, constructed about 400,000 years ago, where domestic fire was used.

The climate they endured was subject to devastating swings between glacial and interglacial periods. At the peak of the Ice Ages, glaciers advanced as far south as England and Germany. Adapting to the harsh environment, *heidelbergensis* developed more muscular bodies with stubby legs and wider chests, better able to conserve heat, slowly evolving into the most famous of early humans, the Neanderthals.

The Neanderthals, the first fossil humans to be discovered, suffered decades of public derision as the result of a description that the French scientist Marcellin Boule gave of a Neanderthal skeleton unearthed near La Chapelle-aux-Saints in 1908. Boule decided that the Neanderthals were little more than savage imbeciles—small-brained, slouching brutes—unable to walk fully upright: 'superior certainly to that of the anthropoid apes, but markedly inferior to that of any modern race whatever'. Boule was confident that they had no direct connection to *Homo sapiens*.

It was not until the 1950s that scientists reached a different verdict. In 1955, after examining the skeleton found at La Chapelle-aux-Saints, two anatomists, William Straus and A.J.E. Cave, published an article pointing out that it belonged to an arthritic 'old man'—an elderly forty-year-old—who had suffered from degenerative joint disease in the skull, jaw, spinal column, hip and feet, as well as a rib fracture and extensive tooth loss. 'Notwithstanding, if he could be reincarnated and placed in a New York subway—provided that he was bathed, shaved, and dressed in modern clothing—it is doubtful whether he would attract any more attention than some of its other denizens'.

Subsequent investigations showed that Neanderthals were highly successful groups of hunter-gatherers, evolving from *heidelbergensis* perhaps as far back as 300,000 years ago, reaching a fully-fledged form by about 120,000 years ago, spreading out across Europe, western Asia and the Levant and surviving through prolonged periods of cold climate. They had stocky, powerful physiques, with large heads, huge noses, low foreheads, double-arched brow ridges, but little chin development. Their brain sizes were similar to those of modern humans but housed in braincases that were elongated from front to back, suggesting that their brains were organised differently. The anatomy of their skulls indicates that they possessed a limited form of vocal communication. They roamed in small nomadic bands over sparsely populated terrain, finding shelter in caves or camping in the open for short periods. They produced a range of specialised tools, referred to as Mousterian, used for hunting, woodwork, and meat and skin preparation. And they were the first human species known to have practiced the simple burial of their dead.

There is also evidence that Neanderthals cared for the infirm and injured. Excavating a cave overlooking the Greater Zab River in the Shanidar Valley in northern Iraq in 1957, an American archaeologist, Ralph Solecki, came across the well-preserved skeleton of an elderly male Neanderthal buried deep beneath the cave floor. A close study of his bones revealed a number of serious injuries: He had suffered a crushing blow to the left side of the head that had fractured the eye socket, displacing the left eye, and probably causing blindness on that side; he had sustained a blow to the right side of the body so severe that the right arm had become withered and useless; and his legs also showed signs of injury. Although disabled and clearly unable to fend for himself, this one-armed, partially blind cripple had managed to survive with his injuries for several years.

Another burial at the Shanidar cave offered an even more tantalising glimpse of Neanderthal life. Soil samples taken by Solecki from the grave of another elderly male showed evidence of large quantities of wildflower pollen. An expert palynologist, Arlette Leroi-Gourhan, claimed that there was far more pollen in the samples than could be accounted for by explanations that it had been blown in on the wind or carried there on the feet of people or animals. The implication, said Solecki, was clear: The elderly man had been buried with an offering of flowers. Writing in *Science* in 1975, at the height of the American hippie movement, Solecki remarked: 'The death had occurred approximately 60,000 years ago . . . yet the evidence of flowers in the grave brings Neanderthals closer to us in spirit than we have ever before suspected . . . The association of flowers with Neanderthals adds a whole new dimension to our knowledge of his humanness, indicating that he had "soul"'. Solecki's verdict was that Neanderthals possessed 'the mind of modern man locked into the body of an archaic creature'. As a title for the book on his work, he chose: *Shanidar: The First Flower People.*

Although many scientists remained sceptical about Solecki's interpretation, the prevailing view during the 1970s was that Neanderthals were advanced enough to be the most likely direct ancestors of *Homo sapiens*.

CHAPTER 17

SAPIENS

TRACING THE ORIGINS of modern humans—*Homo sapiens*—has involved as much controversy as the search for ancient humans. The focus of attention for many decades was on Europe. It was there that the first discoveries were made of the remains of anatomically modern people with an advanced culture dating back 40,000 years that seemed to separate them from all other early human species. In 1868, French workers building a railway at Les Eyzies de Tayac, a village in the valley of the Vézère River in the Dordogne, uncovered the first specimens in a small rock shelter known in the local patois as 'Cro-Magnon'. The name 'Cro-Magnon' became attached not just to the dead occupants buried in the shelter but eventually to the whole population of early modern humans living in Europe during the Upper Palaeolithic period.

The abilities and talents of Europe's Cro-Magnons, when compared to Neanderthals, were prodigious, making them worthy candidates as founder-members of *Homo sapiens*. They ushered in an era of spectacular innovation, producing a range of tools and artefacts far more sophisticated than anything previously found. Their toolkit included long, thin blades of stone, struck off from specially chosen cores and modified further to turn them into specialised knives, scrapers and tools for piercing and engraving. As well as using stone, they worked bone, ivory and antler to manufacture needles, beads and other objects. They also produced composite tools made of several parts, such as harpoons with detachable heads. Most remarkable of all was their artistic endeavour. In what has been described as a 'creative explosion', Cro-Magnons made engravings and sculptures of animals and humans and painted the walls of subterranean caves with vivid images of deer, horses, mammoths, wild cattle and other contemporary beasts.

All this was taken as evidence of a 'human revolution', a flowering of consciousness, which marked the emergence of modern humans. Europe's cave paintings and carved figurines, in particular, were seen as the first stirrings of symbolic and abstract thought and also of language. Murals found in more than 200 caves in western Europe provided testimony of a great artistic outpouring over a period of 25,000 years. The Lascaux caves in the Dordogne, discovered in 1940, contained not only 2,000 figurines but obscure geometric and abstract signs. Bone plaques, found at Les Eyzies and elsewhere, were marked with rows of holes and notches thought to record days, lunar months and the changing seasons. Vulture bones retrieved from cave sites in the Pyrenees Mountains had been turned into flutes with complex sound capabilities. Burial practices had also become more complex, with bodies buried in graves along with goods that would have been considered to be useful in the afterlife. In sum, Europe's Upper Palaeolithic achievements were used as a yardstick for defining modernity.

But a puzzle remained about the sudden appearance in Europe of Cro-Magnons and about who their ancestors were. There were marked physical differences between Cro-Magnons and Neanderthals. The range of their brain size was much the same: between 1,000 and 2,000 cubic centimetres, with an average of 1,350 cubic centimetres. But the Cro-Magnons had higher, more domed skulls, with small brow-ridges and prominent chins; they were taller, with longer legs.

They had linear body shapes like people accustomed to living in warm climates. By contrast, the Neanderthals had shorter, stockier frames, more like people adapted to living in cold climates.

Arguments over the origins of modern humans veered back and forth for much of the twentieth century. During the 1940s, the German-born anatomist Franz Weidenreich challenged Marcellin Boule's verdict that Neanderthals were an evolutionary dead-end and claimed them to be, in fact, the ancestors of modern Europeans. Weidenreich advanced the theory that once *Homo erectus* had emerged from Africa more than 1 million years ago and spread around Eurasia, it had evolved into separate regional groups, each one proceeding in parallel in the same general direction, but developing its own racial characteristics. Thus, China's *Homo erectus* had evolved eventually into modern oriental humans; whereas in Europe, *Homo erectus* had produced the line leading to Neanderthals and modern Europeans.

Little attention was paid at first to Weidenreich's theory of multiregional evolution. The prevailing view remained that Neanderthals were an evolutionary dead-end, replaced some 40,000 years ago by another species, the Cro-Magnons. But during the 1960s, an influential American palaeoanthropologist, Loring Brace, took up the cause of the Neanderthals once again. 'I suggest', Brace wrote in 1964, 'that it was the fate of the Neanderthal to give rise to modern man, and, as frequently happened to members of the older generation in this changing world, to have been perceived in caricature, rejected and disavowed by their own offspring, *Homo sapiens*'.

During the 1970s and early 1980s, another generation of scientists—notably Alan Thorne of the Australian National University and Milford Wolpoff of the University of Michigan—revived the idea of a 'multiregional' model and gained increasing support for it. By the mid-1980s, the majority view was that although the earliest stages of human evolution had occurred in Africa, other key-stages of change and innovation had taken off only after hominids had moved into Eurasia more than 1 million years ago, establishing separate regional groups. The spectacular advances made by Cro-Magnons during Europe's Upper Pleistocene age were a prime example. By comparison, Africa, it was said, had lagged far behind in cultural achievement during the same period, remaining a backwater of little significance.

The growing consensus among palaeoanthropologists about the multiregional origins of modern humans, however, was rudely shattered in 1987 by new findings by molecular biologists. In a paper published by the journal *Nature*, a team from the University of California at Berkeley—Allan Wilson, Rebecca Cane and Mark Stoneking—claimed that genetic evidence showed that 'the modern human family' had originated as a single genetic line in Africa within the last 200,000 years, and not as multiple separate evolutionary events in different parts of the world. Their findings were based on a statistical study of mitochondrial DNA (mtDNA), a special form of DNA that resides not in the nucleus of a cell but in its mitochondria and that is inherited not from both parents but from the mother alone, allowing a trace to be made from living people via generations of mothers into the distant past.

By analysing mtDNA from 147 women from different ethnic groups—Asians, aboriginal Australians, aboriginal New Guineans, Caucasians and 'Africans'—the Berkeley biochemists constructed a genetic tree that identified a common ancestor living in Africa between 142,500 and 285,000 years ago, therefore probably about 200,000 years ago. 'All these mitochondrial DNAs stem from one woman who is postulated to have lived about 200,000 years ago, probably in Africa', they declared. '[A]ll present-day humans are descendants of that African population'.

While the world's media duly celebrated the discovery of an 'African Eve', the scientific community was engulfed in fierce infighting. Palaeoanthropologists in the multiregional camp were furious that upstart biochemists, armed with nothing more than blood samples and computers, had invaded their terrain. 'The fossil record is the real evidence for human evolution', declared Alan Thorne and Milford Wolpoff in an article in *Scientific American*. 'Unlike the genetic data, fossils can be matched to the predictions of theories about the past without relying on a long list of assumptions'.

The multiregionalists pointed out the shortcomings of the Berkeley study. It turned out that of the 147 women who had produced the raw data, ninety-eight had been found in American hospitals; and that of the twenty 'Africans' in the test, only two were actually born in Africa and the eighteen others were African-Americans. Technical aspects of the study were also questioned.

But further research at Berkeley confirmed the initial results. A sample of 189 people, including 121 Africans from six sub-Saharan regions, suggested a common mitochondrial-DNA ancestor in Africa between 166,000 and 249,000 years ago. A Harvard University team, using different mitochondrial tests, produced a date of 220,000 years ago. One study of the complete mitochondrial genome (albeit from only two humans) pointed to an origin date of 176,000 years ago. All the genetic evidence indicated a founding population that lived in Africa in the vicinity of 200,000 years ago. In similar studies, scientists tracking the Y chromosome (a parcel of DNA inherited only through the male line) came up with much the same result.

The fossil evidence to support the theory, however, was sparse. It amounted to little more than odd fragments of bone found in a variety of locations, difficult to date. The most interesting items were finds made in 1967 by Richard Leakey's Omo team when exploring the banks of the Kibish River in southern Ethiopia. These included a partial skull and skeleton known as Omo I and a second skull, Omo II, both dated at about 130,000 years old but clearly identifiable as *Homo sapiens*.

A few researchers during the 1970s argued that Omo I—a male—was a far more likely ancestor for the Cro-Magnons than the Neanderthals. Omo I had a higher and rounder skull and a bigger chin than any Neanderthal, and his skeleton suggested he had a taller and lighter frame. A study by Michael Day and Chris Stringer published in 1982 claimed not only that the Omo skeleton belonged to an early modern human but that it possibly came from the ancestral stock of all living humans. Fossil fragments from sites in South Africa—Klasies River and Border Cave—also pointed to modern-looking humans living there as far back as 100,000 years ago. But otherwise, field evidence to support the idea of an African origin was still meagre.

Two subsequent events dramatically changed the picture. In 1997, members of Tim White's Middle Awash Research Group, exploring a site near the Afar village of Herto, on the Bouri Peninsula, found three fossilised skulls—two adults and a child—which clearly belonged to *Homo sapiens*. Like modern humans, the adults had small faces tucked under capacious braincases, making the facial profile vertical. The cranial volume of one of them was 1,450 cubic centimetres—large even for modern humans. But the skulls did not provide an exact match: They were slightly larger and longer and their brow ridges were more pronounced, echoing features seen in *Homo rhodesiensis/heidelbergensis*. Because of the differences, White's team classified the Herto specimens as a subspecies of *sapiens*, calling it *sapiens idaltu*, an Afar word for 'elder'. White described *idaltu* as 'a population that is on the verge of anatomical modernity but not yet fully modern'. The dates were stunning. All three fossils had been found sandwiched between two volcanic layers that were reliably dated to about 160,000 years ago.

The next revelation came when geologists revisited the site in southern Ethiopia where Richard Leakey's team had found Omo I and II to ascertain a more exact age for the fossils by using advanced dating techniques. Their conclusion, published in *Nature* in 2005, was that the fossils dated back to 195,000 years ago.

Geneticists added their own flourish. A team from the University of Maryland, led by Sarah Tishkoff, reported in 2003 that genetic analysis of more than 600 living Tanzanians showed that they belonged to one of the oldest known human DNA lineages in the world. The Tanzanians came from fourteen different tribes and four linguistic groups; these included the Sandawe, who speak a 'click' language; the Burunge and Gorowaa, who migrated to Tanzania from Ethiopia within the last 5,000 years; and the Maasai and Datog, who probably originated in Sudan. Tishkoff's team chose to investigate East African peoples because the number of linguistic and cultural differences in the region was unusually high and because of the wide variation in the physical appearance of the inhabitants: tall and short; darker-skinned and lighter-skinned; round-faced and narrow-faced. An analysis of the Tanzanians' mitochondrial DNA showed a high degree of genetic variation, or diversity, indicating an ancient lineage: The greater the diversity, the longer a population has existed. Tishkoff's team estimated that the oldest lineages in their study originated some 170,000 years ago.

In further research into Africa's genetic history, published in 2009, Tishkoff's team identified a group with even greater genetic diversity: San hunter-gatherers. Although confined today to areas of southern Africa, the San once occupied much of eastern Africa. Tishkoff concluded that the San represented the oldest population on earth.

Archaeological researchers, too, found increasing evidence that cultural innovation in Africa had occurred far earlier than previously thought. In the 1920s, archaeologists in South Africa had discovered a collection of relatively advanced stone tools—long thin flakes or blades—in a small rock shelter on the side of Howieson's Poort pass near Grahamstown. Other examples of 'Howieson's Poort' tool technology were subsequently found in some twenty rock shelters scattered across southern Africa, indicating that networks of toolmakers had been active. The blades were similar to those found in Europe's Upper Palaeolithic, with an age of 40,000 years or less. But new dating techniques in the 1980s placed the Howieson's Poort industry firmly within Africa's Middle Stone Age, in a period between 80,000 and 60,000 years ago.

Further evidence of advanced toolmaking came from Katanda in the western branch of the Rift Valley in Congo. Exploring sediments in the Semliki River dating back 90,000 years, a team of archaeologists in the 1980s led by John Yellen and Alison Brooks were astonished to find thousands of artefacts that included finely carved harpoons and knives. Local inhabitants had fashioned the harpoons to spear catfish during their spawning season in the shallows, apparently some 50,000 years before Europe's Cro-Magnons had used such sophisticated carving techniques.

During the 1990s, a South African archaeologist, Christopher Henshilwood, uncovered a treasure trove in a cave site he discovered high in a limestone cliff on a wild stretch of the southern Cape coast near Still Bay. The cave entrance at Blombos had been almost totally sealed by dune sand; the cave floor was buried under layers of sediment. But excavations over the years revealed evidence that 75,000 years ago, local inhabitants had been engaged in activities considered to be examples of modern behaviour. They had collected thousands of pieces of ochre from sites at least twelve miles away and turned two of them into tablets engraved with distinct cross-hatched patterns. 'These designs', wrote Henshilwood, 'were engraved with deliberate symbolic intention and had meaning for the maker and very likely for a wider social grouping'. Red ochre pigment, he said, had probably also been used for symbolic body decoration and to colour artefacts.

The Blombos inhabitants had also collected from local river mouths small, pearly white gastropod shells—*Nassarius kraussianus*—turning them into bead 'necklaces' or 'bracelets', providing further evidence of symbolic thinking. 'Each shell was carefully pierced by inserting a small bone tool through the mouth and then with pressure creating a keyhole aperture', said Henshilwood. 'The shells were then strung, perhaps using plant or animal-derived thread, and worn as a personal adornment'.

He maintained that the presence of marine beads, whether used as trade items or to convey group status or to identify group members or relationships within a group, suggested some form of language existed. 'What the beads might symbolize is unknown, but it does imply that there had to be some means of communicating meaning, which plausibly is language'.

Other artefacts found in Blombos Cave from levels dated at about 75,000 years ago included finely worked and polished bone tools—sharp points for hunting weapons and awls for punching holes in tough materials such as animal skins—and leaf-shaped bifacial stone points made in a style previously seen before only in Europe and dated much later at 19,000 years old.

The cave floor also yielded more than 1,000 fish bones, many from large fish—seals and dolphins—and bones from animals such as antelopes. The Blombos community was evidently capable of fishing and hunting large mammals.

While Henshilwood's team was at work at Blombos Cave, another team led by an American palaeoanthropologist, Curtis Marean, began excavations at cave sites on a sandstone cliff at Pinnacle Point, near Mossel Bay, to the east of Still Bay, once sealed off by sand dunes. The discoveries they made there over the course of four years once again changed the narrative of human history. Writing in the journal *Nature* in October 2007, they reported finding three hallmarks of modern life. The inhabitants of Pinnacle Point, they said, had harvested seafood—mussels, clams and snails—and put them over hot rocks to cook; they had manufactured bladelets—tiny blades fashioned from heat-treated silcrete that could be used to form a point for a spear or lined up like barbs on a dart or made into a cutting tool; and they had used pigments—notably red ochre—in ways that appeared to be symbolic, such as body painting. And all this activity was dated as occurring 164,000 years ago.

Marean speculated that the cave dwellers at Pinnacle Point had been in desperate need of new food sources. Africa at the time was experiencing a prolonged cold, dry phase lasting from 195,000 years to 125,000 years ago. The Sahara Desert expanded, virtually cutting off North Africa from the rest of the continent. The Kalahari Desert also expanded, forming another impenetrable barrier. Palaeoenvironmental data suggests that there were only five or six locations in Africa where humans could have survived such harsh conditions. Facing extinction, Marean suggested, inland residents had been forced to search for 'famine food' and migrated to coastal habitats to find it.

'The shellfish may have been crucial to the survival of these early humans', he said. 'Generally speaking, coastal areas were of no use to early humans—unless they knew how to use the sea as a food source. For millions of years, our earliest hunter-gatherer relatives only ate terrestrial plants and animals. Shellfish was one of the last additions to the human diet before domesticated plants and animals were introduced'.

The ability of humans to use food from the sea also meant that they could travel around the coasts of Africa and establish new settlements. 'Coastlines generally make great migration routes. Knowing how to explore the sea for food meant these early humans could now use coastlines as productive home ranges and move long distances'.

He concluded: 'We believe that on the far southern shore of Africa there was a small population of modern humans who struggled through the glacial period 125,000 to 195,000 years ago using shellfish and advanced technologies, and symbolism was important to their social relations. It is possible that this population could be the progenitor population for all modern humans'.

What is notable about the achievements of *Homo sapiens* in this early period in Africa is the pace of innovation. For more than 2 million years, our ancestors had relied on two basic patterns of toolmaking—the Oldowan and the Acheulean traditions—making few changes to the simple stone implements they produced. But from about 300,000 years ago, the archaeological record points to a series of innovations suggesting the dawning of a new age of symbolism and self-awareness. The use of red ochre for body decoration or other purposes is the earliest example. Subsequent innovations included the exploitation of shellfish; the wearing of shell-bead necklaces and bracelets; the manufacture of tiny stone blades potentially hafted for use as weapons; and the trade in materials over distances of 200 miles or more.

From about 100,000 years ago, the pace of innovation accelerated further. New hunting and fishing technologies emerged. Stone points were hafted to make throwing spears, darts and other projectile weapons. Bones were carved into harpoons to facilitate large-scale fishing. Bone awls helped stitch together animal skins for clothing. Ochre plaques were etched with designs suggesting some form of notation. Bow and arrow technology was also developed during this period: researchers exploring Sibudu Cave in northern KwaZulu-Natal, South Africa, have found bone arrowheads dating back to 64,000 years ago, fashioned for use with bows.

Given all this evidence, it now seems certain that by this time African hunter-gatherers had developed a form of articulate language. Scientists remain at odds about the evolution of language: No fossil record exists to help determine how and when vocalisation evolved into language. Some scientists maintain that its slow gestation must have started as far back as 2 million years ago, when early humans needing to cooperate to survive—in foraging for food, for example—turned signals into words. Words eventually gave birth to concepts and proto-language. Other scientists argue in favour of much later development. Paul Mellars, Professor of Pre-History and Human Evolution at Cambridge University, an expert on the evolution of human culture, maintains that it was not until about 100,000 years ago that humans began to fashion 'simple calls' into more complex language.

What is clear is that by 300,000 years ago, African hunter-gatherers had acquired the capacity to create and use symbols. 'Symbolism is the Rubicon that had to be crossed for our ancestors to start becoming human', writes Derek Bickerton in *Adam's Tongue*. The use of symbols—such as body decoration—suggests a faculty for complex communication. Rudimentary language became increasingly needed to transmit information and learned behaviour among individuals and across generations. Indeed, so many hazards did early humans in Africa face that language became an essential tool for survival.

CHAPTER 18

Exodus

Like earlier human relatives, *Homo sapiens* soon ventured out of Africa. During a warm climatic phase starting about 125,000 years ago, *sapiens* migrants moved northwards into the Levant, the lands of the eastern Mediterranean which had a similar environment to their African homeland. Evidence of their existence comes from the remains of individuals buried at cave sites at Jebel Qafzeh and Mugharet es Skhūl, dating back to about 100,000 years. Among the individuals discovered was a man at Skhūl buried with a large pig's jaw in his arms and a child at Qafzeh buried with an antlered deer's skull—the earliest known examples of symbolic burial. But as cold, dry conditions returned, this venture out of Africa eventually failed. By about 90,000 years ago, the Levant migrant community had died out.

The survival of the main African populations of *Homo sapiens* also became more precarious. From about 75,000 years ago, Africa headed into another period of intense cold. Forests shrank; savannahs dried out; the North African desert expanded; sea levels fell to more than 200 feet below their present level. The cold conditions were exacerbated by the eruption of Mount Toba in Sumatra about 74,000 years ago that covered the Indian subcontinent in ash and led to a 'volcanic winter' said to have lasted six years, affecting large parts of the world.

A hypothesis put forward by Stanley Ambrose of the University of Illinois in 1998 argued that the volcanic winter was so severe that it brought about a crash in population levels and caused a 'bottleneck' in human evolution. The effect on the world's climate, according to Ambrose, persisted for 1,000 years. Other scientists doubted that the Toba eruption had such devastating consequences. Nevertheless, harsh conditions in Africa clearly took their toll on *Homo sapiens*. Genetic researchers point to a bottleneck in population numbers some 60,000 years ago. Some estimates suggest the numbers plummeted to as low as 5,000 people.

In dire straits, facing the possibility of extinction, surviving groups were forced to innovate. They began to invent new tools, to form more complex social networks and to become more efficient hunter-gatherers. Some anthropologists speak of 'a great leap forward' in human evolution at this juncture. Richard Klein of Stanford University has argued that a crucial change occurred in human brain capacity, set in motion by a genetic event—'a Big Bang' that heralded the dawn of human culture. Others believe that change accumulated over a more prolonged period, citing the advances made in Africa in fashioning artefacts over a period of 150,000 years. In an article titled 'The Revolution That Wasn't', two of Klein's critics, Sally McBrearty of the University of Connecticut and Alison Brooks of George Washington University, stressed the long-term nature of change. 'As a whole the African archaeological record shows that the transition to fully modern behavior was not the result of biological or cultural revolution, but the fitful expansion of a shared body of knowledge, and the application of novel solutions on an "as-needed basis"'.

What appears certain is that by 60,000 years ago, African hunter-gatherers had developed a fully articulate language, making small groups more cohesive and facilitating long-range planning and the transmission of local knowledge and learned skills.

According to genetic evidence, all human lineages in the world today can be traced back to this ancestral population in Africa. All men alive today carry a Y chromosome which has been passed down from one generation to another and which dates back to this time. Occasional mutations in the Y chromosome down the ages act as genetic markers, enabling researchers to follow the routes taken by modern humans migrating from this ancestral base. Similar information comes from tests of mitochondrial DNA which is passed down the female line without recombining. By using data from both Y chromosomes and mitochondrial DNA, researchers have managed to construct a chain of ancestral clans, or haplogroups—groups of people who share a set of genetic markers and therefore share an ancestor—linking them to past movements.

The most likely location of the homeland of this ancestral population is the northeastern corner of Africa. Population groups found today in Ethiopia—the Oromo and Amhara peoples—retain traces of Y chromosomes that belong to the deepest branch of the Y-chromosome family tree. The same branch of Y chromosomes occurs among San hunter-gatherers who once occupied much of eastern Africa, including Ethiopia. An alternative possibility put forward by Sarah Tishkoff's team based at the University of Maryland is that San migrated into northeastern Africa from what may have been their original homeland in southwestern Africa.

As well as genetic links, there are linguistic clues. The modern San of southern Africa use a variety of 'click' languages involving percussive sounds of considerable complexity. About thirty different click languages are still spoken in southern Africa. The only other place where click languages are spoken is in eastern Africa, among the hunter-gatherers of Tanzania, the Hadzabe and the Sandawe. Recent research led by Stanford's Alec Knight has shown that the San and the Hadzabe belong to different ancient lineages that diverged long ago and that their click languages have no common features other than the use of clicks. The implication is that the language their ancestors used further back in time included clicks and may have formed the basis of a linguistic family stretching from Ethiopia to southern Africa. Knight's team consider it possible that about 60,000 years ago, the use of clicks was widespread before it was filtered out by the evolution of newer languages.

The dispersal of African migrants from their original homeland in northeastern Africa appears to have occurred at some speed. A study of genetic markers shows that the first two branches of Y-chromosome genealogy—haplogroup A and haplogroup B—soon gave rise to sub-branches elsewhere in Africa. Two ancient markers—M 60 and M 91—which date back to about 60,000 years ago are found today in individuals across the continent.

But the most dramatic movement of all was by a small group which left Africa altogether.

By 60,000 years ago, *Homo sapiens* had long been accustomed to a beachcombing lifestyle, exploiting food resources from the sea. During periods of prolonged drought in the interior, Africa's coastal terrain offered a ready refuge. Several sites in South Africa—Klasies River, Blombos Cave and Pinnacle Point—have yielded evidence of seaside living. Another site which gained attention in 2000 is the Abdur reef in Eritrea, on the western shore of the Red Sea. A research team led by an American geophysicist, Robert Walter, reported discovering large middens, or refuse dumps, of shells from clams and oysters, interspersed with obsidian flake tools, dating from about 125,000 years ago. What researchers found of particular interest is that the Abdur reef is in the same neighbourhood as the location from which a group of African emigrants are thought to have crossed from Africa into Arabia.

At the southern end of the Red Sea, as it narrows at the straits of Bab al-Mandab, the distance that separates the coast of Africa from the Arabian peninsula is only fifteen miles. At times of global cooling in the past, when the Red Sea has fallen to levels some 200 feet lower than today, the distance would have been no more than seven miles, with a scattering of islands and reefs offering stepping stones across the straits into new terrain. This crossing point is considered the most likely route taken by Africans expanding into the world beyond.

The group making the crossing according to genetic evidence numbered only a few hundred. One estimate by geneticists in 2005 using mitochondrial DNA data put the figure at most as 550 women of child-bearing age, and probably far fewer. Another estimate suggests a much smaller number, a total of no more than 150 emigrants. What is certain is that the emigrant group contained only a fraction of the full genetic diversity of the existing African population. From the point of their departure, there is a clear divergence between the genetic inheritance of Africa's population and the population of African emigrants who went on to populate the rest of the world.

All women in sub-Saharan Africa belong to one of three branches of the mitochondrial-DNA tree, known as L1, L2 and L3; all women outside Africa belong to two daughter lineages of L3, known as M and N. The likelihood is that a branch of L3 left Africa in a single migration, giving rise to M and N as they made their way eastwards to the Indian subcontinent. A similar pattern emerges from Y-chromosome data. All male lineages outside sub-Saharan Africa carry a Y-chromosome mutation known as M 168.

Most geneticists agree that the crossing took place between 60,000 and 50,000 years ago. A study in 2007 led by Andrea Manica of Cambridge University put forward a date of 55,000 years ago. Generation by generation, the emigrants expanded along the coastlines of southern Arabia to India, leaving behind their genetic footprints. Traces of the mitochondrial M lineage are still commonly found in southern Arabia. Y-chromosome genealogy shows that M 168 soon began to diversify. One branch of emigrants known as M 130 continued their coastal journey eastwards from India, eventually reaching the lost continent of Sunda—the single land mass at the southern end of Asia that once incorporated the Malay Peninsula and the islands of Sumatra, Java and Borneo. Beyond Sunda, some crossed by raft or boat to Australia, then part of another lost continent, Sahul, that included New Guinea and Tasmania. About 60 per cent of Australian aboriginal men have an ancestry directly linked to M 130. Archaeologists estimate that M 130 migrants settled in Australia about 50,000 years ago.

Another branch of M 168 known as M 89 moved into western Asia and the Levant, forming an ancestral base from which other groups migrated into the heart of Eurasia, among them M 9. It was the descendants of M 9 who founded a Y-chromosome lineage of ex-Africans that over the next 30,000 years expanded their range to the furthest ends of the earth.

One group—M 175—moved along a northern route to East Asia, joining migrants who had arrived there earlier along a southern route. Another group—M 45—set up a homeland in the steppes of central Asia, whence their descendants diverged in two directions. One branch headed for Siberia and eventually crossed the Beringia land bridge between Siberia and Alaska that led them to the Americas. Another branch—M 173—headed west into Europe, joining other groups who had migrated there from the Levant.

The advance into Europe of modern hunter-gatherers—Cro-Magnons, as they came to be called—made life even more hazardous for nomadic bands of Neanderthals resident there. Over the past 100,000 years, Neanderthals had managed to survive periods of extreme cold, retreating to refuges in warmer southern areas before returning to the north once the ice sheets receded. But now they faced competition for food and shelter from agile newcomers using sophisticated tools, weapons and language skills and organised in large cohesive groups. As genetic research showed, some interbreeding took place: Traces of Neanderthal genes have been found in all non-African genomes. But as Cro-Magnons moved relentlessly westwards across Europe, Neanderthal numbers steadily dwindled. By about 35,000 years ago, they had been reduced to living in isolated pockets in western Europe. By 30,000 years ago, they had all but vanished.

As they spread out across the world, far-flung branches of the modern human family began to diversify, adapting variously to the different climates and ecologies they encountered. Local populations developed distinctive local features. In northern latitudes, they acquired a lighter skin colour in response to different levels of ultraviolet radiation. The diverse trajectories they followed led to a vast array of languages, cultures and religions. New techniques allowed agriculture and settled communities to flourish. Innovation became a way of life.

Thus it was that a group of African hunter-gatherers led humankind to the threshold of a new world.

Glossary

Absolute dating: The process of determining an approximate age for an archaeological or palaeontological site or artefact, usually based on tests of the physical or chemical properties of materials or other items. Methods include carbon dating and potassium-argon dating. Absolute dating techniques contrast with relative dating techniques, such as stratigraphy.

Acheulean: The name given to an archaeological industry of stone tool manufacture, characterised by large bifaces, including hand-axes. It originated in Africa about 1.5 million years ago, spread to parts of western Asia and Europe and continued in use until about 200,000 years ago.

Adaptive radiation: The rapid diversification of species that occurs after an initial evolutionary innovation.

Archaeology: The study of human behaviour and artefacts in history and prehistory.

Ardipithecus: An early hominid genus found in Ethiopia that lived from about 5.8 million to about 4.4 million years ago.

Australopithecines: A subfamily (Australopithecinae) consisting of a single genus *(Australopithecus)* of extinct hominids that lived from about 4.2 million years ago until about 1 million years ago.

Basalt: A dark, fine-grained volcanic rock.

Biface: A rock core that is flaked on both sides to form a sharp edge around its periphery, such as a hand-axe.

Bipedality: Upright walking on two feet.

Biota: The combined fauna and flora of an area.

Breccia: Rock consisting of angular fragments cemented by finer chalky material. (Italian for 'broken things'.)

Carbon-14 dating: An absolute dating method, based on the decay of the radioactive isotope of carbon, carbon-14.

Chromosomes: Structural elements found in the nucleus of a cell and containing the major part of hereditary material (the genes). Chromosomes are composed of DNA and proteins.

Competitive exclusion principle: The theory that two species cannot exist at the same locality if they have identical ecological requirements.

Continental drift: The movement of continents in geological time due to the drift of the plates of the earth's crust caused by plate tectonics.

Derived character: A new trait developed in a more recent ancestor and retained by descendants but absent in older ancestral stock (which shows a primitive version of the same trait). (See Primitive character.)

DNA: Deoxyribonucleic acid, the molecule carrying hereditary genetic information in all living cells. A DNA molecule consists of a pair of nucleotide chains twisted together in an elegant spiral: the 'double helix'.

Early Stone Age: The first part of the Stone Age, usually applied to Africa, starting around 2.5 million years ago and spanning the Oldowan and Acheulean lithics industries. The same period in Eurasia is referred to as the Lower Palaeolithic.

Fluorine absorption dating: A relative dating technique that determines the duration of time an object has been lying in the soil by measuring the amount of fluorine it has absorbed.

Fossil: The remains or impression of a prehistoric plant or animal that has become hardened into rock.

Hand-axe: A pointed, teardrop-shaped bifacial stone tool most commonly used for butchery purposes.

Haplogroup: A group of people who share a set of genetic markers and therefore share an ancestor. In human genetics, the haplogroups most commonly studied are Y-chromosome haplogroups and mitochondrial-DNA haplogroups, both of which can be used to define genetic populations

Hominid: A term commonly used throughout the period this book covers to refer to all human and pre-human species that ever evolved. In strict taxonomic terms, however, chimpanzees and gorillas are also hominids. Some modern researchers therefore prefer to use the term 'hominin' to describe human and pre-human lineages, thereby excluding chimpanzees and gorillas from the definition. This book keeps to the traditional meaning of hominids. This includes all of the *Homo* species (such as *Homo sapiens*, *Homo ergaster*, *Homo habilis* and *Homo rudolfensis*); all of the australopithecines (such as *Australopithecus africanus* and *Australopithecus afarensis*); and other human ancestors, such as *Ardipithecus*.

Hominin: A term used by modern researchers to describe human and pre-human lineages that excludes chimpanzees and gorillas.

Late Stone Age: The third stage of Africa's Stone Age, starting 50,000 years ago, roughly contemporaneous with Europe's Upper Palaeolithic.

Locum: A doctor standing in for another who is temporarily absent.
Matrix: In palaeontology, the mass of rock or other material in which a fossil is embedded.
Middle Stone Age: The second stage of the Stone Age, applied to Africa, starting about 300,000 years ago, which sees the appearance of more advanced tool-making technologies. The same period in Eurasia is referred to as Middle Palaeolithic.
Miocene: The epoch from 23.3 million to 5.2 million years ago, a period when the first apes appeared and when the common ancestor of humans and chimpanzees split.
Mitochondria: Tiny structures that lie outside the nucleus of a cell and exist as separate organelles with their own DNA. Mitochondrial DNA (mtDNA) is inherited solely through the female line. It is therefore a useful tool for charting population histories.
Molecular clock: The clocklike regularity of the change of a molecule or a whole genotype over geological time.
Morphology: The study of animal structure or form.
Multiregional evolution hypothesis: The hypothesis that modern humans evolved in near concert in different parts of the Old World.
Mutation: An inheritable alteration in genetic material, most commonly an error of replication during cell division.
Natural selection: The process by which in every generation individuals of lesser fitness are removed from the population.
Oldowan: An Early Stone Age industry, first described from Olduvai Gorge, lasting from about 2.6 million to about 1 million years ago. It is characterised by scrapers made from flakes split by hammerstones and by crude choppers made from cobbles—hence the name 'pebble tools'.
Palaeoanthropology: The study of the physical and behavioural aspects of humans in prehistory.
Palaeolithic Age: A prehistoric era in Eurasia divided into three main parts—Lower Palaeolithic, Middle Palaeolithic and Upper Palaeolithic—that are roughly contemporaneous with the three stages of Africa's Stone Age.
Palaeomagnetic dating: A geological technique that dates rock based on the occurrence of polar reversals when the magnetic pole moves to the opposite end of the earth. Such reversals have occurred at regular intervals of hundreds of thousands of years throughout the history of the earth.
Palaeontology: The study of fossils and biology of extinct organisms.
Palynology: The study of contemporary and fossil palynomorphs, such as pollen.
Phyletic gradualism: A mode of evolution characterised by gradual change within a lineage.

Phylogeny: The reconstruction of evolutionary relationships between groups and species, primarily concerned with branching events, usually shown as a branching diagram or phylogenetic tree.

Plate tectonics: The study of the earth's crustal structures, such as continental plates, and the forces that cause them to change shape and move relative to one another.

Pleistocene: The epoch from 1.64 million to about 10,000 years ago, a period which included the Ice Ages and the evolution of the first members of the genus *Homo* in Africa.

Pliocene: The epoch from 5.2 million to 1.64 million years ago when the first hominids evolved in Africa.

Pongidae: The family of great apes that includes chimpanzees, gorillas and orangutans and their ancestors.

Potassium-argon dating: A radiometric dating method based on measurement of the product of the radioactive decay of an isotope of potassium into argon.

Primitive character: A character that was present in a common ancestor of a group and is therefore shared by all members of that group. (See Derived character.)

Proteins: Large molecules that consist of a long chain of amino acids.

Punctuated equilibrium: A mode of evolution characterised by periods of stasis interspersed with brief episodes of rapid change.

Radiometric dating: Clocklike dating methods based on the measurement of the constant rate of decay of naturally occurring radioactive materials.

Relative dating: Techniques that provide information about a site by referring to what is known at other sites or other sources of information, such as faunal correlation. (See Absolute dating.)

Saltation: A sudden event, resulting in a discontinuity, or gap, such as the sudden production of new species.

Speciation: The division of a single parent species into two or more descendant species.

Stasis: A period in the history of organisms during which evolution seemed to have been at a standstill.

Stratigraphy: A branch of geology concerned with the formation, constituents and sequence of stratified deposits.

Tuff: A layer of consolidated volcanic ash and related materials.

Notes on Sources

The material for this book is based largely on the work, writings and reminiscences of several generations of scientists. These chapter notes include references to books I found to be of particular interest and value. A more complete list is contained in the Selected Bibliography.

INTRODUCTION

Since Charles Darwin's time, the study of human evolution has developed into a major industry. Two recent compendiums provide a wealth of information and analysis: *The Complete World of Human Evolution* by Chris Stringer and Peter Andrews (2005); and *Human Evolution: An Illustrated Introduction* by Roger Lewin (2005). In 2007, G. J. Sawyer and Viktor Deak published *The Last Human: A Guide to Twenty-Two Species of Extinct Humans*, which contains remarkable reconstructions of past members of the hominid family. Peter Bowler covers the early development of evolutionary thought in *Evolution: The History of an Idea* (2003). Ian Tattersall, in *The Fossil Trail* (1995), delves into the theoretical disputes among scientists about the course of human evolution, as well as examining the record of fossil discoveries; and in *The World from Beginnings to 4000 BCE* (2008), he presents a sweeping narrative of the major turning points in human evolution. Bernard Wood provides an even more succinct version in *Human Evolution: A Very Short Introduction* (2005).

CHAPTER I

Hans Reck's account of his journey to Olduvai—'The Ravine of Primeval Man'—was first published in 1933. All his original notebooks on Olduvai disappeared during World War I.

Africa's Great Rift Valley is the greatest rupture on the earth's land surface. It was given the name by the English explorer John Gregory in his account of his journey in East Africa in 1893. He first caught sight of the Rift Valley at the Kikuyu Escarpment, just northwest of modern Nairobi. 'We stopped there, lost in admiration of the beauty and in wonder at the character of this valley, until the donkeys threw their loads and bolted down the path'. Part of the Great Rift

Valley in Kenya and northern Tanzania is still known as the Gregory Rift Valley. Nigel Pavitt's book *Africa's Great Rift Valley* (2001) includes stunning photographs.

CHAPTERS 2 AND 3

Raymond Dart held the post of Professor of Anatomy from 1923 to 1958. In an article published in the *Journal of Human Evolution* in 1973, he recalled his horror at the suggestion made by Grafton Elliot Smith that he should apply for the post. 'The very idea revolted me; I turned it down flat instantly. I did not have, as he well knew, the slightest interest in holding a professorship anywhere; least of all one newly-founded, utterly unknown, as remote as possible from libraries and literature and devoid of every other facility for which I had yearned from earliest sentient manhood. He must have been stung by my response, for he said my papers could just as easily be written in the veld! . . .' Advised to consult other colleagues, Dart found them equally in favour of him going. 'Staying would be tantamount to a dereliction of duty . . . Obviously my loss would not be felt in London'.

Dart was remembered by students for his vivid repertoire of lecture-hall tricks. This included leaping up and grasping water pipes attached to the ceiling of the lecture hall to demonstrate the brachiation form of primate locomotion; knuckle-walking like a chimpanzee; and performing a 'crocodile walk' to illustrate how reptiles moved about. A former pupil, Trevor Jones, described him as 'a brilliant lecturer, a superb actor with a mischievous streak'. But he was also remembered for his notorious temper, earning the nickname 'The Terror of the Dissection Hall'. 'He would stride between the tables watching', recalled Trevor Jones, 'and very often would jump onto a table to give expression to his thoughts and critics in the best of Australian'.

As a result of his nervous breakdown, he stepped down from his post for a year in 1943. In postwar years, he worked on a treasure trove of fossilised broken bones found in limestone caves in the Makapansgat Valley in the northern Transvaal, developing a theory that australopithecines had been 'flesh hunters' who used bones as weapons to attack and kill their own kind. In a landmark paper headed 'The Predatory Transition from Ape to Man', published in 1953, Dart wrote with his customary flair of how australopithecines had 'seized living quarries by violence, battered them to death, tore apart their broken bodies, dismembered them limb from limb, slaking their ravenous thirst with the hot blood of victims and greedily devouring livid writhing flesh'.

Humankind, according to Dart, had descended from this 'killer ape', inheriting the same habit of using violence and weapons. 'The loathsome cruelty of mankind to man forms one of his inescapable, characteristic and differentiative features; and it is explicable only in terms of his carnivorous, and cannibalistic origin', he wrote. 'The blood-spattered, slaughter-gutted archives of human history from the earliest Egyptian and Sumerian records to the most recent atrocities of the Second World War accord with early universal cannibalism, with animal and human sacrificial practices, or their substitutes in formalized religions, and with the world-wide scalping, head-hunting, body mutilating and necrophiliac practices of mankind proclaiming this common bloodlust differentiator, this predacious habit, this mark of Cain that separates man dietetically from his anthropoid relatives and allies him rather with the deadliest of carnivores!'

Dart's young colleague Phillip Tobias tackled him about his tendency to 'overdo the text'. Tobias recalled: 'He looked at me, not unkindly but bridling a little and said, "Phillip, I have to do it this way, with such a new and revolutionary concept," then added: "If you don't give them [expletive] 200 per cent, they [expletive]-well won't believe the half of it"'.

Dart's 'killer-ape' hypothesis was greeted largely by scepticism from the scientific community. Nor did he win much support for his claim that he had discovered 'a new age of man'—a 'Bone Age' which he called the 'Osteo-donto-keratic', a culture based on the use of bones, tooth and horn. Nevertheless, Dart's version of humankind's bloody ancestry soon became part of modern folklore. The American dramatist Robert Ardrey picked up the idea and turned it into a highly successful book, *African Genesis*, published in 1961.

Subsequent studies of South Africa's cave systems carried out by Dr Bob Brain of the Transvaal Museum revealed a different picture from the one Dart painted. Whereas Dart had argued that the fossil bones at Makapansgat showed evidence of bloodthirsty fighting between killer apes with weapons made of bone, Brain made a convincing case that cave-site bones had been accumulated by animal predators and scavengers such as hyenas, leopards and sabre-toothed cats, and that rather than being the hunters, our apelike ancestors were more likely to have been the hunted.

In 1966, the *South African Journal of Science* published a special edition to commemorate the centenary of Robert Broom's birth. It included accounts by his son, Norman Broom, and by Professor Lawrence Wells. Wells referred to Broom's combative nature and recalled that in 1910 'an unfortunate train of events' had led him to throw over his post of Professor of Zoology and Geology at Victoria College. 'It seems likely that then as in later years he would never have stepped an inch out of his way to avoid a fight, but rather would have taken a perverse delight in steering a collision course'.

CHAPTERS 4, 5 AND 6

Robert Broom's verdict on Louis Leakey was: 'He has the restlessness of the true hunter, always looking for something new; and with the intuition of true genius generally looking in the right spot'.

Both Louis Leakey and Mary Leakey have provided valuable autobiographical accounts of their endeavours. Describing the view from the rim of Ngorongoro that she came to know so well, Mary Leakey wrote: 'As one comes over the shoulder of the volcanic highlands to start the steep descent, so suddenly one sees the Serengeti, the plains stretching away to the horizon like the sea, a green vastness in the rains, golden at other times of the year, fading to blue and grey. Away to the right are the Precambrian outcrops and an almost moonlike landscape. To the left, the great slopes of the extinct volcano Lemagrut dominate the scene, and in the foreground is a broken, rugged country of volcanic rocks and flat-topped acacias, falling steeply to the plains. Out on the plains can be seen small hills—like Naibor Soit, Engelosen or Kelogi near Olduvai: the scale is so vast that one cannot tell that the biggest is several hundred feet high. Olduvai Gorge can also be seen. Two narrow converging dark lines, softened by distance and heat haze, pick out the Main Gorge and Side Gorge, each of which is in reality many miles long and in places half a mile across'.

Virginia Morell's comprehensive biography of the Leakey family provides many candid insights.

Efforts to identify the culprits involved in the Piltdown forgery still continue, more than fifty years after Joseph Weiner, Kenneth Oakley and Wilfred Le Gros Clark exposed it in 1953 : 'The solution of the Piltdown problem', *Bulletin of the British Museum of Natural History, Geology* 2 (3) : 139–146 (1953). Brian Gardiner restated the case against Martin Hinton in 2003. Hinton died in 1961. Charles Dawson died in 1916.

CHAPTER 7

Richard Leakey's output of publications was as prolific as his father's. His early exploits were all the more remarkable because for ten years he suffered from a debilitating kidney disease. The disease was diagnosed in 1968, but it was not until 1979, after his brother Philip offered to donate one of his kidneys, that he underwent a successful transplant. In 1989, he left the field of palaeoanthropology to take up an appointment as head of the Kenyan government's wildlife department, with the mandate to root out corruption and poaching. The poaching menace was significantly reduced, but in the process Leakey offended many local politicians and officials. (See *Wildlife Wars: My Battle to Save Kenya's Elephants*, Macmillan, London [2001]). In 1993, he lost both his legs after a plane crash. Sabotage was suspected but never proved. In 1995, he helped launch an opposition party. After pressure from international donor organisations, President Daniel Arap Moi appointed him cabinet secretary and head of the civil service in 1997. He resigned in 2001.

Roger Lewin deals with the KBS tuff controversy in *Bones of Contention* (1997). The complex stratigraphy of the Turkana Basin was eventually sorted out by Frank

Brown, an American geologist who made an extensive study of it in the 1980s. Brown describes tuffs as 'thin slices of time'. After each volcanic eruption, ashes that ride on the wind accumulate on the landscape, forming layers that range from a few inches thick to fifty feet. In some cases—as at Olduvai—the tuff consists of a relatively orderly layer cake which can be dated without much difficulty. But geologists investigating the tuff at East Turkana found 'a mess'. Ashfalls there—numbering in all about seventy—were mixed with volcanic ash brought down by rivers and streams from the Ethiopian highlands. Brown solved the problem by using a new technique that enabled each tuff to be identified by its unique geochemical 'fingerprint'.

Virginia Morell examines in detail Richard Leakey's relationships with Donald Johanson and Tim White. Delta Willis writes vividly about Leakey's exploits in *The Hominid Gang* (1989). Meave Leakey's comments about Richard are taken from an interview she gave in June 2004, published by the Academy of Achievement, Washington, DC.

CHAPTERS 8, 9, 10 AND 11

Lewis Nesbitt, or Ludovico Mariano Nesbitt, as he was otherwise known, was an Italian-born mining engineer of British descent. The U.S. edition of his account of his Afar journey was called: *Hell-Hole of Creation: The Exploration of the Abyssinian Danakil* (1934).

Maurice Taieb was the first scientist to discover the importance of Hadar, but he soon found himself shunted aside by Donald Johanson, who spoke openly of his determination to become as famous as Richard Leakey. In an interview with Virginia Morell in 1986, Taieb recalled that one day Johanson had asked him whether he thought it possible for him to become as famous as Richard Leakey, as he had neither a famous mother, nor a famous father, nor a famous family name. Taieb remarked: 'I think he decided that since he could not be a Leakey, he had to beat them; he had to be against them'. Indeed, the dust jacket of Johanson's book *Lucy* (1981) boasts of how he had mounted the 'first real and successful challenge to the Leakey dynasty'.

As a graduate student, Tim White spent three seasons from 1974 to 1976 working at Richard Leakey's Turkana base. In 1975, with Richard Leakey's encouragement, he also began making a detailed study of Mary Leakey's Laetoli fossils. But he fell out with Richard Leakey in 1976 during the KBS tuff controversy. White accused Leakey of trying to censor the outcome of his own research because it did not support Leakey's side of the argument. Along with Johanson, he became an outspoken critic of the Leakeys.

The feuds of this period are covered in detail by Virginia Morell, Roger Lewin, Jon Kalb, and Delta Willis. Alan Walker writes about the discovery of Turkana Boy in *The Wisdom of the Bones* (1996).

CHAPTER 12

Elisabeth Vrba, a graduate in zoology and statistics from the University of Cape Town, decided in 1968 to spice up her life as a high-school teacher in Pretoria by applying for voluntary work at the Transvaal Museum. The director, Bob Brain, an expert on South African cave fossils, gave her a pile of rocks containing antelope fossils to clean and sort. From such humble beginnings, Vrba developed an expertise on antelope species that led her to challenge standard assumptions about their evolutionary history.

Hitherto, biologists had measured evolutionary 'success' in terms of the number of species that a particular animal had produced. Hence, wildebeest, which had split into some forty different species during the previous 6 million years, were regarded as more 'successful' than impala, which had produced only one or two species. A study that Vrba conducted in Kruger National Park revealed a different perspective. Vrba noted that not only were impala there far more numerous than all the wildebeest and other antelopes combined but that their lifestyles were different. As generalists, impalas thrived in a variety of habitats, ranging from savannah to woodland, consumed most kinds of vegetation, and saw little need to migrate. Wildebeest, by contrast, were specialists, preferring to graze in dry, open savannah and willing to migrate long distances in search of a suitable niche. Specialists, she concluded, were more affected by environmental change and thus more prone to evolutionary pressures than generalists.

Vrba's 1980 paper 'Evolution, Species and Fossils' gained international attention. As deputy director of the Transvaal Museum, she also became drawn into hominid studies, observing that dramatic episodes of evolutionary change in antelope history—such as occurred 2.5 million years ago—appeared to coincide with crucial events in hominid history. The common cause, she concluded, was climate change. 'We palaeontologists saw evidence long before the climatologists found it of massive climatic change 2.5 million years ago by looking in the fossil record'.

CHAPTER 13

In *The First Human* (2006), Ann Gibbons follows the exploits of four teams of scientists as they raced to discover the earliest ancestor: Tim White and members of the Middle Awash Research Group in Ethiopia; Meave Leakey's team at Kanapoi in Kenya; Michel Brunet's Mission Paléoanthropologique Franco-Tchadienne; and Martin Pickford and Brigitte Senut, co-leaders of the Kenya Paleontology Expedition. Among Meave Leakey's team was her daughter, Louise, a skilled palaeontologist from the third generation of Leakeys to take to the field.

In 1999, an Ethiopian expedition leader, Zeresenay Alemseged, exploring the Dikika area adjacent to Hadar, discovered the earliest and most complete *afaren-*

sis juvenile yet found. He spent five years separating bones from a block of sandstone 'grain by grain'. He named the juvenile 'Selam', meaning 'peace'.

CHAPTER 14

In 1999, the fossil sites of Sterkfontein, Swartkrans, Kromdraai and the surrounding area—popularly known as 'the Cradle of Humankind'—were declared a World Heritage Site. About 40 percent of the world's human ancestor fossils have been found there. Opening a visitor centre in 2005, President Thabo Mbeki remarked: 'I would ask you to be very still. If we are very still, we will hear, if we really listen, these rocks and stones speaking to us today. They are the voices of our distant ancestors, who still lie buried in them'. A collection of essays entitled *Origins* (2006), edited by Geoffrey Blundell, includes contributions from Ron Clarke, Christopher Henshilwood on Blombos Cave, and Ben Smith on the rock arts of sub-Saharan Africa.

CHAPTERS 15, 16, 17 AND 18

The initial discovery of stone tool sites in the Hadar-Gona River area was made in the 1970s by Gudrun Corvinus, a German archaeologist, who was a member of Maurice Taieb's expedition. Her work was followed by Hélène Roche, a French archaeologist, and then by an international team led by Sileshi Semaw, who carried out fieldwork in Gona between 1992 and 1994. Semaw's team confirmed a date for the oldest tools found there as being 2.6–2.5 million years old. 'The artefacts show surprisingly sophisticated control of stone fracture mechanics', they reported in *Nature* in 1997.

The 'tool factory' at Kenya's Lokalalei site was found by a team led by Anne Delagnes and Hélène Roche. They surmised that the tools were probably used partly to prepare food. Remains of ancient cattle, pigs, horses, rhinos and even crocodiles were found at the site. 'The hominids likely lived in small groups exploring their environment for food procurement, either hunting or scavenging small game, or collecting fruits and plants', said Delagnes.

The hobbit was discovered by a team led by Michael Morwood, an Australian archaeologist who tells the tale in *A New Human* (2007). An Australian anatomist, Peter Brown, studied the bones for three months before reaching a verdict in 2004. The announcement of the discovery triggered a worldwide debate. Analysing the foot bones, William Jungers of Stony Brook University, New York, suggested in 2009 that the ancestors of *Homo floresiensis* may not have been *Homo erectus* but 'some other, more primitive hominid whose dispersal into southeast Asia is still undocumented' (*Nature* 459 : 81–84 [2009]).

Another puzzling discovery was made in 2008 by a team of Russian researchers excavating a rock shelter at Denisova in the Altai Mountains in Siberia. The cave site had previously yielded a collection of tools left behind by Neanderthals who had lived there between 48,000 and 30,000 years ago. Modern humans—Cro-Magnons—were also known to have lived in the same region at the same time. When the researchers dug up a section of finger bone at Denisova, they assumed that it belonged either to one of the Neanderthal inhabitants or to a modern human. But a small sample of mitochondrial DNA material extracted from the bone fragment told a different story: It matched that of neither Neanderthals nor modern humans.

Research results published in 2010 (*Nature* 468, 1012) showed that it belonged to a young girl from a new hominid lineage that researchers called Denisovan. Their conclusion was that Neanderthals and Denisovans were cousins, sharing a common ancestor that left Africa half a million years ago. While the Neanderthals spread westwards towards Europe, the Denisovans headed eastwards, inhabiting Siberia as recently as 30,000 years ago. Some Denisovans had interbred with *sapiens* groups migrating eastwards from Africa about 50,000 years ago: Genetic analysis of present-day occupants of Melanesia (Papua New Guinea and islands northeast of Australia) indicated they have inherited about 5 percent of their DNA from Denisovan roots.

Scientists are divided over whether there was a single migration out of Africa or a multiple exodus. Two Cambridge researchers, Robert Foley and Marta Lahr, argue in favour of at least two waves, an early one via the southern route to India and later migrations moving northwards via Suez and the Levant to Europe and Asia. Stephen Oppenheimer, an Oxford-based researcher, puts the case for a single exodus via southern Arabia. They also differ over the date of the first exodus. Oppenheimer places it before the Toba eruption 74,000 years ago. The Cambridge school believe that the first migration occurred nearer to 60,000 years ago. Spencer Wells describes the endeavours of geneticists to decipher the genetic code of modern humans. Nicholas Wade provides a masterful summary of recent research.

The fate of the Neanderthals is examined by Christopher Stringer and Clive Gamble, by Erik Trinkhaus and Pat Shipman, and by Ian Tattersall. In 2010, an international research team studying the Neanderthal genome reported finding convincing evidence of limited interbreeding between Neanderthals and *Homo sapiens*. 'We found the genetic signal of Neanderthals in all the non-African genomes', said Ed Green of the University of California at Santa Cruz, the lead author of the study. The research team estimated that about 2 per cent of the genomes of present-day humans living from Europe to Asia—and as far into the Pacific Ocean as Papua New Guinea—was inherited from Neanderthals. This suggested that interbreeding had occurred soon after the initial migration of *Homo sapiens* out of Africa when they first encountered Neanderthals, probably in the Middle East. No trace of Neanderthals' DNA was found in present-day Africans.

Selected Bibliography

Alemseged, Z., *et al.* 'A Juvenile Early Hominid Skeleton from Dikika, Ethiopia', *Nature* 443 : 296–301 (2006)

Ambrose, S. 'Late Pleistocene Human Population Bottlenecks, Volcanic Winter, and Differentiation of Modern Humans', *Journal of Human Evolution* 34 (6) : 623–651 (1998)

Ardrey, R. *African Genesis: A Personal Investigation into the Animal Origins and Nature of Man*, Collins, London (1961)

Asfaw, B., *et al.* '*Australopithecus garhi*: A New Species of Early Hominid from Ethiopia', *Science* 284 : 629–635 (1999)

Asfaw, B., *et al.* 'Remains of *Homo erectus* from Bouri, Middle Awash, Ethiopia', *Nature* 416 : 317–320 (2002)

Backwell, L., *et al.* 'Middle Stone Age Bone Tools from the Howiesons Poort Layers, Sibudu Cave, South Africa', *Journal of Archaeological Science* 35 : 1566–1580 (2010)

Balter, M. 'Paleontological Rift in the Rift Valley', *Science* 292 : 198–201 (2001)

Beard, C. *The Hunt for the Dawn Monkey: Unearthing the Origins of Monkeys, Apes and Humans*, University of California Press, Berkeley (2006)

Behar, D.M., *et al.* 'The Dawn of Human Matrilineal Diversity', *American Journal of Human Genetics* 82 : 1130–1140 (2008)

Berger, L., *et al.* '*Australopithecus sediba*: A New Species of *Homo*-like Australopith from South Africa', *Science* 328 : 195–204 (2010)

Berger, L., and Hilton-Barber, B. *In the Footsteps of Eve: The Mystery of Human Origins*, Adventure Press, National Geographic, Washington, DC (2000)

Bickerton, Derek. *Adam's Tongue: How Humans Made Language; How Language Made Humans*, Hill and Wang, New York (2009)

Blundell, G. (ed.). *Origins: The Story of the Emergence of Humans and Humanity in Africa,* Double Storey, Cape Town (2006)

Bonnefille, R. 'Evidence for a Cooler and Drier Climate in the Ethiopian Highlands Towards 2.5 Myr Ago', *Nature* 303 : 487–491 (1983)

Bonner, P., *et al.* (eds.). *A Search for Origins: Science, History and South Africa's 'Cradle of Humankind',* Wits University Press, Johannesburg (2001)

Botha, R., and Knight, C. (eds.). *The Cradle of Language*, Oxford University Press, Oxford (2009)

Bowler, P.J. *Evolution: The History of an Idea*, third edition, University of California Press, Berkeley (2003)
———. *Life's Splendid Drama; Evolutionary Biology and the Reconstruction of Life's Ancestry 1860–1940*, University of Chicago Press (1996)
———. *Theories of Human Evolution: A Century of Debate, 1844–1944*, Johns Hopkins University Press, Baltimore (1986)
Brace, C.L. 'The fate of the "Classic" Neanderthals: A Consideration of Hominid Catastrophism', *Current Anthropology* 5 : 3–43 (1964)
———. *The Stages of Human Evolution: Human and Cultural Origins*, Prentice Hall, Englewood Cliffs, NJ (1967)
Brain, C.K. *The Hunters or the Hunted? An Introduction to African Cave Taphonomy*, University of Chicago Press (1981)
Brain, C.K., and Sillen, A. 'Evidence from the Swartkrans Cave for the Earliest Use of Fire', *Nature* 336 : 464–466 (1988)
Brooks, A.S., *et al.* 'Dating and Context of Three Middle Stone Age Sites with Bone Points in the Upper Semliki Valley, Zaire', *Science* 268 : 548–553 (1995)
Broom, N. 'Address on the Opening of the Robert Broom Museum, 1 December 1966', *South African Journal of Science* 63 (9) : 371–377 (1967)
Broom, R. 'Discovery of a New Skull of the South African Ape-Man, *Plesianthropus*', *Nature* 159 : 672 (1947)
———. 'A New Fossil Anthropoid Skull from South Africa', *Nature* 138 : 486–488 (1936)
———. *The Search for Man's Ancestry*, Watts, London (1950)
———. 'The Sterkfontein Ape', *Nature* 139 : 326 (1937)
Broom, R., and Robinson, J.T. 'A New Type of Fossil Man', *Nature*, 164 : 322–323 (1949)
Broom, R., and Schepers, G. *The South African Fossil Ape-Men; the Australopithecinae*, Transvaal Museum Memoir No. 2, Pretoria (1946)
Brown, P., *et al.* 'A New Small-Bodied Hominin from the Late Pleistocene of Flores, Indonesia', *Nature* 431 : 1055–1061 (2004)
Brunet, M. 'Discovery in Chad, Central Africa; Towards a New Paradigm for the Cradle of Mankind', in *Origins*, ed. G. Blundell, Double Storey, Cape Town (2006)
Brunet, M., *et al.* '*Australopithecus bahrelghazali*, une nouvelle espèce d'Hominidé ancien de la région de Koro Toro (Tchad)', *Comptes Rendus de l'Académie des Sciences*, 324 : 341–345 (1996)
———. 'The First Australopithecine 2,500 km West of the Rift Valley (Chad)', *Nature* 378 : 273–275 (1995)
———. 'A New Hominid from the Upper Miocene of Chad, Central Africa', *Nature* 418 : 145–151 (2002)
Burton, Frances D. *Fire: The Spark That Ignited Human Evolution*, University of New Mexico Press, Albuquerque (2009)
Butler, D. 'The Battle of Tugen Hills', *Nature* 410 : 508–509 (2001)

Cann, R., Stoneking, M., and Wilson, A. 'Mitochondrial DNA and Human Evolution', *Nature* 325 : 31–36 (1987)

Cavalli-Sforza, L. *Genes, Peoples and Languages*, Penguin, London (2000)

Clarke, R. 'Discovery of Complete Arm and Hand of the 3.3 Million-Year-Old *Australopithecus* Skeleton from Sterkfontein', *South African Journal of Science* 95 : 477–480 (1999)

———. 'Dr Broom and the Skeleton in the Cavern', in *Origins*, ed. G. Blundell, Double Storey, Cape Town (2006)

———. 'First Ever Discovery of a Well-Preserved Skull and Associated Skeleton of *Australopithecus*', *South African Journal of Science* 94 : 460–463 (1998)

Cole, S. *Leakey's Luck: The Life of Louis Leakey, 1903–1972*, Collins, London (1975)

Coppens, Y. 'East Side Story: The Origin of Humankind', *Scientific American* 270 (5) : 88–95 (1994)

Dart, R. 'Africa, the Cradle of Humanity', *Illustrated London News*, June 13, 1925.

———. '*Australopithecus africanus*: The Man-Ape of South Africa', *Nature* 115 : 195–199 (1925)

———. 'Recollections of a Reluctant Anthropologist', *Journal of Human Evolution* 2 : 417–427 (1973)

———. 'Robert Broom—His Life and Work', *South African Journal of Science* 48 (1) : 3–19 (1951)

Dart, R., and Craig, D. *Adventures with the Missing Link*, Hamilton, London (1959)

Darwin, C. *The Descent of Man, and Selection in Relation to Sex*, Murray, London (1871)

———. *On the Origin of Species by Means of Natural Selection*, Murray, London (1859)

Dawkins, R. *The Ancestor's Tale; A Pilgrimage to the Dawn of Life*, Weidenfeld & Nicolson, London (2004)

Day, M. '*Early Homo sapiens* Remains from the Omo River Region of South-West Ethiopia. Omo Human Skeletal Remains', *Nature* 222 : 1135–1138 (1969)

Day, M., Leakey, M.D., and Olson, T. 'On the Status of *Australopithecus afarensis*', *Science* 207 : 1102–1103 (1980)

Day, M., and Napier, J.R. 'Hominid Fossils from Bed I, Olduvai Gorge, Tanganyika. Fossil Foot Bones', *Nature* 201 : 967–970 (1964)

Deacon, H.J., and Deacon J. *Human Beginnings in South Africa; Uncovering the Secrets of the Stone Age,* Philip, Cape Town (1999)

Delagnes, A., and Roche, H. 'Late Pliocene Hominid Knapping Skills: The Case of Lokalalei 2C, West Turkana, Kenya', *Journal of Human Evolution* 48 (5) : 435–472 (2005)

d'Errico, F., *et al.* 'Archaeological Evidence for the Emergence of Language, Symbolism, and Music—An Alternative Multidisciplinary Perspective', *Journal of World Prehistory* 17 : 1–70 (2003)

———. '*Nassarius kraussianus* Shell Beads from Blombos Cave: Evidence for Symbolic Behaviour in the Middle Stone Age', *Journal of Human Evolution* 48 : 3–24 (2005)

DeSalle, R., and Tattersall, I. *Human Origins: From Bones to Genomes*, Texas A & M Press, College Station (2007)

DeSilva, J. 'Functional Morphology of the Ankle and the Likelihood of Climbing in Early Hominins', *Proceedings of the National Academy of Sciences* 106 : 6567–6572 (2009)

Diamond, J. *Guns, Germs and Steel; The Fates of Human Societies*, Norton, New York (1998)

———. *The Third Chimpanzee; The Evolution and Future of the Human Animal*, Harper, New York (1992)

Dugard, J. 'Palaeontologist Ron Clarke and the Discovery of "Little Foot": A Contemporary History', *South African Journal of Science* 91 : 563–566 (1995)

Eldredge, N., and Gould, S.J. 'Punctuated Equilibria: An Alternative to Phyletic Reconstruction', in T.J.M. Schopt and J.M. Thomas (eds.), *Models in Paleobiology*, Freeman, San Francisco (1972)

Esterhuysen, A. *Sterkfontein; Early Hominid Site in the 'Cradle of Humankind'*, Wits University Press, Johannesburg (2001)

Findley, G. *Dr Robert Broom*, F.R.S. Balkema, Cape Town (1972)

Foley, R. 'Adaptive Radiations and Dispersals in Hominin Evolutionary Ecology', *Evolutionary Anthropology* 11 : 32–37 (2002)

———. *Humans Before Humanity*, Blackwell, Oxford (1995)

———. 'Speciation, Extinction and Climatic Change in Hominid Evolution', *Journal of Human Evolution* 26 : 275–289 (1994)

Gardner, B. 'The Piltdown Forgery: A Re-Statement of the Case Against Hinton', *Zoological Journal of the Linnean Society* 139 : 315–335 (2003)

Gardner, B. and Currant, A. *The Piltdown Hoax. Who Done It?* Linnean Society of London (1996)

Gibbons, A. *The First Human: The Race to Discover Our Earliest Ancestors*, Doubleday, New York (2006)

Green, R.E., *et al.* 'A Draft Sequence of the Neanderthal Genome', *Science* 326 : 710–722 (2010)

Gregory, J.W. *The Great Rift Valley, being the Narrative of a Journey to Mount Kenya and Lake Baringo with some account of the Geology, Natural History, Anthropology and Future Prospects of British East Africa*, Murray, London (1896); republished by Cass, London (1968)

Gregory, W.K., and Hellman, M. 'Evidence of the Australopithecine Man-Apes on the Origin of Man', *Science* 88 : 615–616 (1938)

Gundling, T. *First in Line: Tracing Our Ape Ancestry*, Yale University Press, New Haven (2005)

Haeckel, E. *Natürliche Schöpfungsgeschichte* (1868), published as *The History of Creation*, London (1876)

Haile-Selassie, Y. 'Late Miocene Hominids from the Middle Awash, Ethiopia', *Nature* 412 : 178–181 (2001)

Haile-Selassie, Y., *et al.* 'An Early *Australopithecus afarensis* Postcranium from Woranso-Mille, Ethiopia', *Proceedings of the National Academy of Sciences* 107 : 12,121–12,126 (2010)

———. 'Late Miocene Teeth from Middle Awash, Ethiopia and Early Hominid Dental Evolution', *Science* 303 : 1503–1505 (2004)

Hamilton, A. *Environmental History of East Africa*, Academic Press, London (1982)

Henshilwood, C. 'Modern Humans and Symbolic Behaviour; Evidence from Blombos Cave, South Africa', in G. Blundell (ed.), *Origins*, Double Storey, Cape Town (2006)

Henshilwood, C., *et al.* 'An Early Bone Tool Industry from the Middle Stone Age at Blombos Cave, South Africa: Implications for the Origins of Modern Human Behaviour, Symbolism and Language', *Journal of Human Evolution* 41 : 631–678 (2001)

———. 'Emergence of Modern Human Behaviour: Middle Stone Age Engravings from South Africa', *Science* 295 : 1278–1280 (2002)

———. 'Middle Stone Age Shell Beads from South Africa', *Science* 304 : 404 (2004)

Hilton-Barber, B., and Berger, L. *The Official Field Guide to the Cradle of Mankind; Sterkfontein, Swartkrans, Kromdraai & Environs World Heritage Site*, Struik, Cape Town (2002)

Hopwood, A. 'Miocene Primates from Kenya', *Journal of the Linnean Society London* 38 : 437–464 (1933)

Johanson, D. 'Lucy (*Australopithecus afarensis*)', in M. Ruse and J. Travis (eds.), *Evolution: The First Four Billion Years*, Belknap, Cambridge, MA (2009)

Johanson, D., and Edey, M. *Lucy: The Beginnings of Humankind*, Simon & Schuster, New York (1981)

Johanson, D., and Edgar, B. *From Lucy to Language,* Simon & Schuster, New York, revised edition (2006)

Johanson, D., *et al.* 'New Partial Skeleton of *Homo habilis* from Olduvai Gorge, Tanzania', *Nature* 327 : 205–209 (1987)

Johanson, D., and Shreeve, J. *Lucy's Child: The Discovery of a Human Ancestor*, Morrow, New York (1989)

Johanson, D., and White, T. 'On the Status of *Australopithecus afarensis*', *Science*, 207 : 1104–1105 (1980)

———. 'A Systematic Assessment of Early African Hominids', *Science* 203 : 321–330 (1979)

Johanson, D., White T., and Coppens, Y. 'A New Species of the Genus *Australopithecus* (Primates: Hominidae) from the Pliocene of Eastern Africa', *Kirtlandia* 28 : 1–14 (1978)

Johanson, D., and Wong, K. *Lucy's Legacy: The Quest for Human Origins*, Harmony, New York (2009)
Jones, S., Martin, R., and Pilbeam, D. *Cambridge Encyclopedia of Human Evolution*, Cambridge University Press (1992)
Kalb, J. *Adventures in the Bone Trade: The Race to Discover Human Ancestors in Ethiopia's Afar Depression*, Copernicus, New York (2001)
Keith, A. *An Autobiography*, Watts, London (1950)
———. *New Discoveries Relating to the Antiquity of Man*, Williams and Norgate, London (1931)
———. *A New Theory of Human Evolution*, Watts, London (1948)
Keith, A., *et al.* 'The Fossil Anthropoid from Taungs', *Nature* 115 : 234–236 (1925)
Kingdon, J. *Lowly Origins: Where, When and Why Our Ancestors First Stood Up*, Princeton University Press (2003)
Klein, R. *The Human Career: Human Biological and Cultural Origins*, second edition, Chicago University Press (1999)
Klein, R., and Edgar B. *The Dawn of Human Culture: A Bold New Theory on What Sparked the "Big Bang" of Human Consciousness*, Wiley, New York (2002)
Knight, A., *et al.* 'African Y Chromosome and mtDNA Divergence Provides Insight into the History of Click Languages', *Current Biology* 13 : 464–473 (2003)
Krause, J., *et al.* 'The Complete Mitochondrial DNA Genome of an Unknown Hominin from Southern Siberia,' *Nature* 464 : 472–473 (2010)
Lahr, M. 'The Multiregional Model of Modern Human Origins', *Journal of Human Evolution* 26 : 23–56 (1994)
Lahr, M. and Foley, R. 'Multiple Dispersals and Modern Human Origins', *Evolutionary Anthropology* 3 : 48–60 (1994)
———. 'Towards a Theory of Modern Human Origins: Geography, Demography and Diversity in Recent Human Evolution', *Evolutionary Anthropology* 41 : 137–176 (1998)
Leakey, L.S.B. *Adam's Ancestors*, fourth edition, Methuen, London (1953)
———. *By the Evidence: Memoirs, 1932–1951*, Harcourt Brace Jovanovich, New York (1974)
———. 'Exploring 1,750,000 Years into Man's Past', *National Geographic*, October 1961, 120/4 : 564–589
———. 'Finding the World's Earliest Man', *National Geographic*, September 1960, 118/3 : 420–435
———. 'Fossil Human Remains from Kanam and Kanjera, Kenya Colony', *Nature* 138 : 643 (1936)
———. 'New Finds at Olduvai Gorge', *Nature* 189 : 649–650 (1961)
———. 'A New Fossil Skull from Olduvai', *Nature* 201 : 967–970 (1959)
———. *The Stone Age Cultures of Kenya Colony*, Cambridge University Press (1931)

———. *The Stone Age Races of Kenya*, Oxford University Press, Oxford (1935)
———. *White African*, Hodder and Stoughton, London (1937)
Leakey, L.S.B., Evernden, J.F., and Curtis, G. 'Age of Bed I, Olduvai Gorge, Tanganyika', *Nature* 191 : 478–479 (1961)
Leakey, L.S.B., *et al.* 'Age of the Oldoway Bone Beds, Tanganyika', *Nature* 128 : 724 (1931)
Leakey, L.S.B., Tobias, P.V., and Napier, J.R. 'A New Species of the Genus *Homo* from Olduvai Gorge, Tanzania', *Nature* 202 : 308–312 (1964)
Leakey, M.D. *Disclosing the Past; An Autobiography*, Weidenfeld and Nicolson, London, (1984)
———. *Olduvai Gorge: My Search for Early Man*, Collins, London (1979)
Leakey, M.D., and Hay, R. 'Pliocene Footprints in the Laetoli Beds at Laetoli, Northern Tanzania', *Nature* 278 : 317–323 (1979)
Leakey, M.G., and Harris, J. (eds.). *Lothagam: The Dawn of Humanity in Eastern Africa*, Columbia University Press, New York (2003)
Leakey, M.G., *et al.* 'New Four Million Year Hominid Species from Kanapoi and Allia Bay, Kenya', *Nature* 376 : 565–571 (1995)
———. 'New Hominin Genus from Eastern Africa Shows Diverse Middle Pliocene Lineages', *Nature* 410 : 419–420 (2001)
Leakey, R. 'Early *Homo sapiens* Remains from the Omo River Region of South-West Ethiopia. Faunal Remains from the Omo Valley', *Nature* 222 : 1132–1133 (1969)
———. *The Making of Mankind*, Joseph, London (1981)
———. *One Life: An Autobiography*, Joseph, London (1983)
Leakey, R., and Lewin, R. *Origins: What New Discoveries Reveal About the Emergence and Evolution of Our Species and Its Possible Future*, Macdonald and Jane's, London (1977)
———. *Origins Reconsidered: In Search of What Makes Us Human*, Doubleday, New York (1992)
———. *People of the Lake: Man, His Origins, Nature and Future*, Collins, London (1979)
Lewin, R. *Bones of Contention: Controversies in the Search for Human Origins*, second edition, University of Chicago Press (1997)
———. *Human Evolution; An Illustrated Introduction*, fifth edition, Blackwell, Oxford (2005)
Lombard, M., and Phillipson, L. 'Indications of Bow and Stone-Tipped Arrow Use 64,000 Years Ago in KwaZulu-Natal', South Africa, *Antiquity* 84 : 635–648 (2010)
Lovejoy, O. 'Evolution of Human Walking', *Scientific American* 259 : 118–125 (1988)
Macaulay, V., *et al.* 'Single, Rapid Coastal Settlement of Asia Revealed by Analysis of Complete Mitochondrial Genomes', *Science* 308 : 1034–1036 (2005)
Marean, C. 'When the Sea Saved Humanity', *Scientific American* 303 : 55–61 (2010)

Marean, C., *et al.* 'Early Human Use of Marine Resources and Pigment in South Africa During the Middle Pleistocene', *Nature* 449 : 905–909 (2007)

Maslin, A.M., and Beth Christensen. 'Tectonics, Orbital Forcing, Global Climate Change, and Human Evolution in Africa', *Journal of Human Evolution* 53 : 443–464 (2007)

Mayr, E. *What Evolution Is*, Weidenfeld & Nicolson, London (2002)

McBrearty, S., and Brooks, A.S. 'The Revolution That Wasn't: A New Interpretation of the Origin of Modern Human Behaviour', *Journal of Human Evolution* 39 : 453–563 (2000)

McDougall, I., *et al.* 'Stratigraphic Placement and Age of Modern Humans from Kibish, Ethiopia', *Nature* 433 : 733–736 (2005)

McPherron, S.P., *et al.* 'Evidence for Stone-Tool-Assisted Consumption of Animal Tissues Before 3.39 Million Years Ago at Dikika, Ethiopia', *Nature* 466 : 857–860 (2010)

Morell, V. *Ancestral Passions: The Leakey Family and the Quest for Humankind's Beginnings*, Simon & Schuster, New York (1995)

Morwood, M., and Oosterzee, P. *A New Human: The Startling Discovery and Strange Story of the "Hobbits" of Flores, Indonesia*, Collins, London (2007)

Murray, B. *Wits, The Early Years: A History of the University of the Witwatersrand Johannesburg and Its Precursors, 1896–1939*, Witwatersrand University Press, Johannesburg (1982)

Nesbitt, L. *Desert and Forest: The Exploration of the Abyssinian Danakil*, Cape, London (1934)

Olsen, S. *Mapping Human History: Discovering the Past Through Our Genes*, Houghton Mifflin, Boston (2002)

Oppenheimer, S. *Out of Eden; The Peopling of the World*, Constable, London (2003)

Palmer, D. *Seven Million Years: The Story of Human Evolution,* Weidenfeld & Nicolson, London (2005)

Parfitt, S.A., *et al.* 'Early Pleistocene Human Occupation at the Edge of the Boreal Zone in Northwest Europe', *Nature* 466 : 229–233 (2010)

Pavitt, N. *Africa's Great Rift Valley*, Abrams, New York (2001)

Pickford, M., and Senut, B. 'The Geological and Faunal Context of Late Miocene Hominid Remains from Lukeino, Kenya', *Comptes Rendus de l'Académie des Sciences* 332 (2) : 145–152 (2001)

———. ' "Millennium Ancestor", a 6-Million-Year-Old Bipedal Hominid from Kenya', *South African Journal of Science* 97 (1–2) : 22 (2001)

Reader, J. *Missing Links: The Hunt for Earliest Man*, second edition, Pelican, London (1988)

Reck, H. *Die Schlucht des Urmenschen*, Brockhaus, Leipzig (1951)

Reed, K. 'Paleoecological Patterns at the Hadar Hominin Site, Afar Regional State, Ethiopia', *Journal of Human Evolution* 54 : 743–768 (2008)

Richmond, B., and Jungers, W. '*Orrorin tugenensis* Femoral Morphology and the Evolution of Hominin Bipedalism', *Science* 319 : 1662–1665 (2008)

Roche, H., *et al.* 'Early Hominid Stone Tool Production and Technical Skill 2.34 Myr Ago in West Turkana, Kenya', *Nature* 399 : 57–60 (1999)

Russell, M. *Piltdown Man: The Secret Life of Charles Dawson and the World's Greatest Archaeological Hoax*, Tempus, Stroud (2003)

Sarich, V., and Wilson A. 'Immunological Time Scale for Hominid Evolution', *Science* 158 : 1200–1203 (1967)

Sawyer, G., and Deak, V. (eds.). *The Last Human: A Guide to Twenty-two Species of Extinct Humans*, Nèvraumont, New York (2007)

Scholz, C., *et al.* 'East African Megadroughts Between 135 and 75 Thousand Years Ago and Bearing on Early-Modern Human Origins', *Proceedings of the National Academy of Sciences* 104 : 16,416–16,421 (2007)

Schwartz, J. *Sudden Origins: Fossils, Genes, and the Emergence of Species*, Wiley, New York (1999)

Semaw, S., *et al.* 'The world's Oldest Stone Artefacts from Gona, Ethiopia: Their Implications for Understanding Stone Technology and Patterns of Human Behaviour Between 2.6–1.5 Million Years Ago', *Journal of Archaeological Science* 27 : 1197–1214 (2000)

———. '2.6-Million-Year-Old Stone Tools and Associated Bones from OGS-6 and OGS-7, Gona, Afar, Ethiopia', *Journal of Human Evolution* 45 : 169–177 (2003)

Semino, O., *et al.* 'Ethiopians and Khoisan Share the Deepest Clades of the Human Y-Chromosome Phylogeny', *American Journal of Human Genetics* 70 : 265–268 (2002)

Senut, B., *et al.* 'First Hominid from the Miocene (Lukeino Formation, Kenya)', *Comptes Rendus de l'Académie des Sciences* 332 (2) : 137–144 (2001)

Shipman, P. *The Man Who Found the Missing Link; Eugène Dubois and his Lifelong Quest to Prove Darwin Right*, Simon & Schuster, New York (2001)

Simpson, G.G. *Tempo and Mode in Evolution*, Columbia University Press (1944)

Simpson, S., *et al.* 'A Female *Homo erectus* Pelvis from Gona, Ethiopia', *Science* 322 : 1089 (2008)

Solecki, R. *Shanidar: The First Flower People*, Knopf, New York (1971)

———. 'Shanidar IV, a Neanderthal Flower Burial in Northern Iraq', *Science* 190 : 880–881 (1975)

Soodyall, H. (ed.). *The Prehistory of Africa: Tracing the Lineage of Modern Man*, Ball, Johannesburg (2006)

Spencer, F. *Piltdown: A Scientific Forgery*, Natural History Museum/Oxford University Press, Oxford (1990)

Sperber, G. *From Apes to Angels: Essays in Anthropology in Honor of Phillip V. Tobias*, Wiley, New York (1990)

Stanley, S. *Children of the Ice Age: How a Global Catastrophe Allowed Humans to Evolve*, Harmony, New York (1996)

Straus, W., and Cave, A. 'Pathology and Posture of Neanderthal Man', *Quarterly Review of Biology* 32 (4) : 348–363 (1957)

Stringer, C. 'Coasting out of Africa', *Nature* 405 : 24–27 (2000)

———. 'Modern Human Origins: Progress and Prospects', *Philosophical Transactions of the Royal Society of London* 357B : 563–579 (2002)

Stringer, C. and Andrews, P. *The Complete World of Human Evolution*, Thames & Hudson, London (2005)

Stringer, C., and Gamble, C. *In Search of the Neanderthal; Solving the Puzzle of Human Origins,* Thames & Hudson, London (1993)

Stringer, C. and McKie, R. *African Exodus; The Origins of Modern Humanity*, Cape, London, 1996

Swisher C.S., Curtis G.H., and Lewin, R. *Java Man: How Two Geologists' Dramatic Discoveries Changed Our Understanding of the Evolutionary Path to Modern Humans*, Scribner, New York (2001)

Taieb, T. *Sur la terre des premiers hommes*, Laffont, Paris (1985)

Tattersall, I. *Becoming Human: Evolution and Human Uniqueness*, Oxford University Press, Oxford (1999)

———. *The Fossil Trail; How We Know What We Think We Know About Human Evolution,* Oxford University Press, New York (1995)

———. *The Last Neanderthal: The Rise, Success and Mysterious Extinction of Our Closest Human Relatives*, Westview, Boulder (1998)

———. *The Monkey in the Mirror: Essays on the Science of What Makes Us Human*, Oxford University Press, Oxford (2002)

———. *The World from Beginnings to 4000 BCE*, Oxford University Press, New York (2008)

Tattersall, I., and Schwartz, J. *Extinct Humans*, Westview, Boulder (2000)

Taylor, T. *The Artificial Ape: How Technology Changed the Course of Human Evolution*, Palgrave Macmillan, New York (2010)

Terry, R. '*Raymond Dart: Taung 1924–1974',* 50th Anniversary Commemoration Booklet, Museum of Man and Science, Johannesburg (1974)

Thesiger, W. 'The Awash River and the Aussa Sultanate', *Geographical Journal of London* 85 : 1–23 (1935)

———. *The Life of My Choice*, Collins, London (1987)

Theunissen, B. *Eugène Dubois and the Ape-Man from Java: The History of the First 'Missing Link' and Its Discoverer*, Kluwer, Dordrecht (1989)

Thorne, A. and Wolpoff, M. 'The Multiregional Evolution of Modern Humans', *Scientific American* 266 : 76–83 (1992)

———. 'Regional Continuity in Australasian Pleistocene Hominid Evolution', *American Journal of Physical Anthropology* 55 : 337–350 (1981)

Thorpe, S., Crompton, R., and Holder, R. 'Origin of Human Bipedalism as an Adaptation for Locomotion on Flexible Branches', *Science* 316 : 1328 (2007)

Tishkoff, S. and Verrelli, B. 'Patterns of Human Genetic Diversity: Implications for Human Evolutionary History and Disease', *Annual Review of Genomic and Human Genetics* 4 : 293–340 (2003)

Tishkoff, S., and Williams, S. 'Genetic Analysis of African Populations: Human Evolution and Complex Disease', *Nature Reviews: Genetics* 3 : 611–621 (2002)

Tishkoff, S., *et al.* 'The Genetic Structure and History of Africans and African Americans', *Sciencexpress*, April 30, 2009

Tobias, P.V. *Dart, Taung and the 'Missing Link',* Witwatersrand University Press, Johannesburg (1984)

———. *Into the Past: A Memoir*, Picador Africa, Johannesburg (2005)

Trinkhaus E., and Shipman P. *The Neandertals: Changing the Image of Mankind*, Knopf, New York (1992)

Underhill, P., *et al.* 'The Phylogeography of Y Chromosome Binary Haplotypes and the Origins of Modern Human Populations', *Annals of Human Genetics* 65 : 43–62 (2001)

Vigilant, L., *et al.* 'African Populations and the Evolution of Human Mitochondrial DNA', *Science* 253 : 1503–1507 (1991)

Vrba, E.S. 'Chronological and Ecological Implications of the Fossil Bovidae at the Sterkfontein Australopithecine Site', *Nature* 250 : 19–23 (1974)

———. 'Ecological and Adaptive Change Associated with Early Hominid Evolution', in F. Delson (ed.), *Ancestors: The Hard Evidence*, Liss, New York (1985)

———. 'Evolution, Species and Fossils: How Does Life Evolve?' *South African Journal of Science* 76 : 61–84 (1980)

———. 'Late Pliocene Climatic Events and Hominid Evolution', in F. Grine (ed.), *Evolutionary History of the 'Robust' Australopithecines,* Aldine de Gruyter, New York (1988)

———. 'Some Evidence of the Chronology and Palaeoecology of Sterkfontein, Swartkrans, and Kromdraai from the Fossil Bovidae', *Nature* 254 : 301–304 (1975)

Vrba, E.S., *et al. Paleoclimate and Evolution with Emphasis on Human Origins*, Yale University Press, New Haven (1995)

Wade, N. *Before the Dawn: Recovering the Lost History of Our Ancestors*, Penguin, New York (2006)

Walker, A., *et al.* '2.5 Myr *Australopithecus boisei* from West of Lake Turkana, Kenya', *Nature* 322 : 517–522 (1986)

Walker, A., and Shipman, P. *The Ape in the Tree; An Intellectual and Natural History of Proconsul*, Belnap Press, Cambridge, MA (2005)

———. *The Wisdom of the Bones; In Search of Human Origins*, Weidenfeld and Nicolson, London (1996)

Walter, R.C., *et al.* 'Early Human Occupation of the Red Sea Coast of Eritrea During the Last Interglacial', *Nature* 405 : 65–69 (2000)

Watson, E., *et al.* 'Mitochondrial Footprints of Human Expansion in Africa', *American Journal of Human Genetics* 61 : 691–704 (1997)

Weidenreich, F. 'Facts and Speculations Concerning the Origin of *Homo sapiens*', *American Anthropologist* 49 : 187–203 (1947)

———. 'The "Neanderthal Man" and the ancestors of "*Homo sapiens*"', *American Anthropologist* 45 : 39–48 (1943)

Weiner, J. *The Piltdown Forgery*, Oxford University Press, Oxford (1955)

Wells, L.H. 'One Hundred Years: Robert Broom, 30 November 1866–6 April 1951,' *South African Journal of Science* 63 (9) : 357–366 (1967)

Wells, S. *Deep Ancestry; Inside the Genographic Project*, National Geographic Society, Washington, DC (2007)

———. *The Journey of Man: A Genetic Odyssey*, Princeton University Press (2002)

Wheelhouse, F., and Smithford, K. *Dart: Scientist and Man of Grit,* Transpareon Press, Sydney (2001)

White, T., *et al.* '*Ardipithecus ramidus* and the Paleobiology of Early Hominids', *Science* 326 (5049) : 75–86 (2009)

———. '*Australopithecus ramidus*, a New Species of Early Hominid from Aramis, Ethiopia', *Nature* 371 : 306–312 (1994)

———. 'Pleistocene *Homo sapiens* from Middle Awash, Ethiopia', *Nature* 423 : 742–747 (2003)

Willis, D. *The Hominid Gang: Behind the Scene in the Search for Human Origins*, Penguin, New York (1989)

———. *The Leakey Family: Leaders in the Search for Human Origins*, Facts on File, New York (1992)

Willoughby, P. *The Evolution of Modern Humans in Africa*, Altamira, Lanham (2007)

WoldeGabriel, G., *et al.* 'Ecological and Temporal Placement of Early Pliocene Hominids at Aramis, Ethiopia', *Nature* 371 : 330–333 (1994)

———. 'Geology and Palaeontology of the Late Miocene Middle Awash, Afar Rift, Ethiopia', *Nature* 412 : 175–178 (2001)

Wolpoff, M., *et al.* 'Modern *Homo sapiens* Origins: A General Theory of Hominid Evolution Involving Evidence from East Asia', in F. Smith and F. Spencer (eds.), *The Origins of Modern Humans: A World Survey of the Fossil Evidence*, Liss, New York (1984)

———. 'Multiregional, Not Multiple Origins', *American Journal of Physical Anthropology* 112 : 129–136 (2000)

Wood, B. *Human Evolution: A Very Short Introduction,* Oxford University Press, Oxford (2005)

Wrangham, R. *Catching Fire: How Cooking Made Us Human*, Profile Books, London (2009)

Yellen, J.E., *et al.* 'A Middle Stone Age Worked Bone Industry from Katanda, Upper Semliki Valley, Zaire', *Science* 268 : 553–556 (1995)

Zimmer, C. *Where Did We Come From? The Essential Guide to Human Evolution*, Apple Press, Hove, England (2006)

Index

Index

Index

Martin Meredith is a journalist, biographer and historian who has written extensively on Africa and its recent history. He is the author of many books, including *Diamonds, Gold and War: The Making of South Africa*; *The State of Africa: A History of the Continent Since Independence*; *Mugabe: Power, Plunder and the Struggle for Zimbabwe*; *Coming to Terms: South Africa's Search for Truth*; and *Mandela: A Biography.* He lives near Oxford, England.